AF541078

ETHNOBIOLOGY :
Therapeutics and Natural Resources

ETHNOBIOLOGY : Therapeutics and Natural Resources

by

Dr. Ashis Kumar Ghosh
M.Sc., M.Ed., Ph.D., Diploma in Naturopathy
H.M., Habibpur Saraswati Vidyamandir
Midnapore, Paschim Medinipur-721101 (W.B.)

2009

DAYA PUBLISHING HOUSE
DELHI-110 035

ISBN 81-7035-576-1
ISBN 978-81-7035-576-2

Published by : **Daya Publishing House**
1123/74, Deva Ram Park
Tri Nagar, Delhi - 110 035
Phone: 27383999
Fax: (011) 23260116
E-mail: dayabooks@vsnl.com
Website: www.dayabooks.com

Showroom : 4760-61/23, Ansari Road,
New Delhi-110 002
Ph.: 23245578, 23244987

Laser Composed by : **Vaishnav Graphics & Systems**
New Delhi-110 055

Printed at : **Chawla Offset Printers**
Delhi - 110 052

PRINTED IN INDIA

Dedication

The author dedicates this work to the memory of Late Abinash Chandra Ghosh (great grandfather, former headmaster during 1906-1914 of Rajagram S.B. Raha Institution, P.O.-Rahagram, Bankura which is previously known as Rajagram A.S. School) and to late grandfather Bhutnath Ghosh, starter of Calcutta Tram Company and to my parents.

Dr. Dulal Chandra Pal
M.Sc., Ph.D (Cal), F.B.S., F.E.S., F.I.A.T. M.N., A.Sc.
Secretary, Gram Seva Sangha
P.O. – Hatthuba, North 24 Parganas, W.B.
Pin-743269

FOREWORD

It is known that there is a tremendous progress in development of modern medicine. Still today, medicinal plants continue to be an important source of drugs throughout the world. In fact in developed countries, there is a perceptible revival of the use of herbs in health care programmes. It is probably due to continues introduction of a large number of synthetic drugs and antibiotics by pharmaceutical companies resulted in wide spread Toxicity. Today more than 25 percent of the formulations contain plant products. These has also been considerable increase in the sale of herbal teas and health foods. People in general are now prefer to use natural products.

Plants as source of medicine are much more important for developing countries like India. About 85 percent of the people are still dependent on traditional systems of medicine, which derives more than 90 percent of the medicaments from higher plants.

In this perspective, unwritten and undocated tribal system of medicine has tremendous potentialities in new sources of raw materials for development of new drugs both for human and veterinary care.

Practically there is no book on this subject hence the present book entitled "Ethnobiology : Therapeutics and Natural Resources" written by Dr. Ashis Kumar Ghosh will be a valuable tool in the hand of students and scientists in the areas of botany, pharmacology, chemistry and pharmaceutical industries.

Dr. Dulal Chandra Pal
Retd. Scientist – I,
Govt. of India

ACKNOWLEDGEMENTS

I am privileged to record my gratitude to Prof. Sabyasachi Chatterjee (Kalantoo), Joydev De (Asst. Director), Dr. Goutam Bhattacharya; S.U.; Dr. Debal Deb; Dr. Satyahari De, IIT, Kharagpur; Prof. D.P. Kushari, B.U., for their keen interest in this work. My sincere thank is also due to my colleagues and M.C. Members of Saraswati Vidyamandir for their encouragement. In the preparation of this book I have received valuable suggestions and generous help from the following persons.

Dr. Madhup De, A.I. (S.E.), Paschim Medinipur; Syed Nurus Salam, D.I. (S.E.), Paschim Medinipur; Shankari Harh, D.P.O., Bankura; Dr. Subhasis C. Patra, Biman Gupta, Joint Secretary, District Library, Paschim Medinipur; Sudip Bej, Demonstrator, District Computer Centre, Paschim Medinipur; Aparesh Bhattacharya, District Secretary, ABTA, Medinipur; Dr. D.C. Pal, Sr. Scientist, B.S.I. Calcutta; Prof. S. D. Mishra, BARC, Mumbai; Ajoy Sen & Asim Chakraborty (D.I. Office) Paschim Medinipur; Prof. Subhendu Mukherjee, C.U.; Prof. A.K. Biswas, Bishnupur; Tara Shankar Biswas, Working Secretary (Edn.), Paschim Medinipur; Nisith Das, Secretary of Sabyasachi Patrika, Paschim Medinipur.

Author wants to express his thanks to Sarathi Maity (S.I.), Biplab Ghosh and Soma Ghosh for the patience shown by them during the preparation of the manuscript. Thank is also due to all the staff of D.I. Office for their encouragement.

Dr. Ashis Kumar Ghosh

PREFACE

It is being increasingly realized that "Green Medicine" plays a vital role in our life. We do not usually remember that we get the most important things in our life like plants to breathe, food and medicine as well as numerous gifts of nature, without which our life would have been impossible, free of cost. Indiscriminate exploitation of such natural resources is creating problems which pose various threat to living beings. So, it is essential that we should learn from our student life how to conserve, nurture and nourish the green medicine alongwith the environment. In this book the author has done a commendable effort and highlighted the green medicine for high tech low cost solution.

Though developed countries have solved their immediate health problems to a large extent, developing countries are yet to come up. In order to tackle such problems India Govt. have introduced National Green Corps (N.G.C.) from school level for general awareness creation to university level. The word Ethnobotany was coined about 100 years ago but it became more widely known only through N.G.C. as green medicine during last one decade. The green medicine refers to the subject of total relationship between man, animal and plant life. Books, research papers or general articles under the title "Green medicine" have been rather few. It is a green pharmacy consisting new discoveries in herbal remedies for common diseases alongwith harmful plants, human beliefs and the acupressure. Still thousands of safe, natural remedies lie untapped in forests and rural areas throughout the India. In this book you may share knowledge of healing power about the following ailments:-Abscesses, Ageing, Allergies, Angina,

Arthritis, Asthma, Athlete's Foot, Bad Breath, Baldness, Breast feeding problem, Bronchitis, Burns, Cancer prevention, Cataracts, Chronic Fatigues Syndrome, Cold and Flu, Constipation, Corns, Coughing, Cuts, Dandruff, Depression, Diabetes, Diarrhoea, Dizziness, Erection problems, Fainting, Fever, Flatulence, Fungal infection, Gall and kidney stones, High blood pressure, High Cholesterol, Hypo and Hyper thyroidism, Indigestion, Infertility, Insect Bite and Stings, Insomnia, Intestinal Parasites, Lice, Liver problems, Menopause, Pregnancy and Delivery, Toothache, T.B., Ulcers etc.

The research work incorporated in the book has been carried out after proper investigation during the period from 1993 to 2003 by Dr. Ashis Ghosh and the co-workers. Some of the works has already published in reputed Journals and books of India and abroad.

The book will be a good companion and a helpful guide to students, research scholars, teachers, myth makers, folk doctors, pharmaceutist, ayurvedic and homeopathic doctors even also interested lay public.

No inconsistencies might have crept in.

Rathayatra, **Dr. Ashis Kumar Ghosh**

CONTENTS

Foreword .... (vii)
Acknowledgement .... (ix)
Preface .... (xi)

1. INTRODUCTION **1-2**

2. BENEFICIAL **3-59**

2.1 For Human **3**

2.1.1 Herbal Medicine from Mangrove Areas of South 24 Pgs. 3

2.1.2 An Ethno-Medico-Ecological Survey at Paschim Medinipur District for Natural Health Care and Green Belt Movement 3

2.1.3 A Contribution to the Ethnobotany of Gujarat: Some Medicinal Plants Used as Antidiabetic, Antipyretic and Icteric Agents 14

2.1.4 Fruit Can Play as an Antidote in Different Ailments 18

2.1.5 Instant Household Remedies for Dyspepsia and Cooking Industry 20

2.1.6 Potential Remedies for Everyday 21

2.1.7 How You Can Survive for a Long Time in the Earth? 25

2.1.8 Herbal Folk Remedies of Bankura and Medinipur Districts, West Bengal 33

2.2 **For Animals** 37

2.2.1 Ethnoveterinary Medicines from the Tribal Areas of Bankura and Medinipur Districts, West Bengal 37

2.2.2 Effects of an Aqueous Extract of *Cardiospermum halicacabum* Linn. Seed on Experimentally Induced Rat Paw Oedema 41

2.2.3 Herbal Veterinary Medicine from the Tribal Areas of Midnapur & Bankura Districts, West Bengal 44

2.2.4 Indigenous Household Human-Animal Health Care Practices Through Ethnomedicinal Animals 47

3. **ENVIRONMENTAL AWARENESS** **61-170**

3.1 Traditional Vegetable Dyes from Central West Bengal 61

3.2 Natural Vegetable Dyes for Textile Industry in South Bengal 65

3.3 Effects of Polluted Air of Durgapur in Some Medicinal Plants 67

3.4 Selected Piscicidal Plants from South Bengal to Catch Fish 70

3.5 Beware of the Selected Harmful Plants to Keep the Body Fit 77

3.6 Acupressure – A Way of Natural Living 84

3.7 Let's Go to Know Our Popular Superstitions, Magico-religious Beliefs and on Some Curious Cults on Plants, Animals and Events 92

3.8 Beneficial Common Birds in Bengal 97

3.9 Panther's Appeal to Human to Protect Biodiversity 104

3.10 Sacred Grove Relics as Bird Refugia 105

3.11 Foot Prints of Mammals 110

3.12 Apiphilic Plants in Agro-forestry for Better Bee Management to Get More Honey and Wax 111

3.13 Clay Dyes of Purba Medinipur – A Gift of Nature 113

3.14 Some Valuable Aspects on Bio-natural Cultivation 115

3.15 Pest Management Practices Through Natural Resources 116
3.16 Non-Conventional or Additional as well as Natural Source of Different Economic Products from Palms in South Bengal 118
3.17 Ethno-Nutritious Self-Made Baby Food from Paschim Medinipur 120
3.18 The Natural Method of Healing of Ethnic People Through Soil Therapy and Hydrotherapy 121
3.19 Natural Method of Healing of Ethnic Groups Through Common Practices and Dietotherapy 125
3.20 Apiphilic Plants in Agro-Forestry for Bee Management 130
3.21 Medicinal Plants Used for Treatment of Diabetes by the Tribal Peoples of Bankura, Purulia and Medinipur of West Bengal 132
3.22 Cows (*Bos indicus*) Make Medicine for Mankind 140
3.23 Pest Management Practices Through Biocides 142
3.24 Indigenous Method for Lustry Building by Natural Components 143
3.25 Ecophilic Fuels for Rail Engines and Automobiles 145
3.26 Wood-Ash – The Natural Preserver 146
3.27 Natural Way of Sugarcane Cultivation Through Dew-Trapping Process 147
3.28 Tailoring of Haulms Serve as a Natural Remedy Against "Late Blight of Potato" and Natural Resource for Bio-Fertilizer 149
3.29 The Naturally Cultivated and Processed Groundnut Milk is the Alternative of Cow Milk 150
3.30 Seriphilic Plants in Agro-Forestry 152
3.31 The New Sweety Natural Safe Beverage Yielding Plant of Bengal 153
3.32 Hitherto Untapped Plantlore from Bengal 155
3.33 Sapling of Bamboos Through Nodal Cuttings in Bengal 163
3.34 Senescence Retardant and Antioxidant Potential Plants 165

3.35 Natural Dye-Yielding Plants from West Bengal 167
3.36 Effect of Natural Bio-Fertilizer on Kendu (*Diospyros exsculpta*) Leaves 169

4. GLOSSARY FOR MEDICAL TERMS **171-181**

5. REFERENCES **182-188**

6. LIST OF PUBLICATIONS **189-192**

7. INDEX **193-204**

7.1 Plant Index 193
7.2 Animal Index 198
7.3 Disease Index 200

1
INTRODUCTION

A vast knowledge on Green medicine exists as oral among the folklore and primitive societies which are still much to be discovered. The indigenous systems of medicine in our country have deep roots in our cultural heritage. From the immemorial, in order to bring relief from pain and diseases several Green medicine were in use. This system is mainly based on the drug therapy from natural resources. These references are incorporated in ancient literature, namely Atharvaveda, Charak Samhita, Sushruta Samhita, Astanga – Hridaya etc. The drug industry in India has developed considerably since 1947. On the whole, the industry had to depend largely on import of diverse chemicals. Today the Green medicine has to cope with increasing demand for genuine and reliable medicine and in an effort for mass production. The more important one among the many problems is the procurement of right kind and quality crude drugs in the required quantity. Agro-forestry can also meet this demand.

In earlier days the Green medicine practitioner used to prepare medicines for the patients himself with full sense of devotion and sincerity but with the increasing civilization led to the tendency towards more and more dependence on ready made. There is a need to develop gene bank of tested and certified germplasms. The germplasm should be properly monitored to retain the inherent Green medicine. The tissue culture techniques can reduce the time period in rearing the Green medicine. It is essential because the continuous and irregular exploitation of Green medicines wealth of forests is likely to deteriorate their relative supply position for Industrial utilization and deplete our forests of the valuable genetic stock. Besides home garden can be introduced which is a land use form on private land surrounding individual houses with a definite fence in which several tree (cult alongwith medicinal) species are cultivated together with flower and fruit yielding plants according to the individual needs of the family to save the eco-system. Finally it can be claimed the Nation can be judged by its activity towards plants and animals.

2
BENEFICIAL

2.1 FOR HUMAN

2.1.1 Herbal Medicince from Mangrove Areas of South 24 Pgs., West Bengal

The major mangrove areas in India are the Gangetic Sunderbans Complex. The Gangetic Sunderbans (W. B.) lies under the East Coast. Sunderbans is the largest mangrove belt of the World, most of which falls in Bangladesh. It is generally considered that Sunderbans derive their name from the word Sundari, the local name for the tree, *Heritiera fomes.*

Rhizophora apiculata (Garjan), *R. mucronata* (Garjan), *Ceriops decandra* (Goran), *Bruguiera cylindrica* (Sona champa), *B. gymnorrhiza* (Kakra), *B. parviflora* (Champa), *Xylocarpus granatum* (Dhundul) are the dominant species here. *Porteresia coarctata* (Dhani), a saline grass related to rice abundantly grows on newly formed areas, *Nypa fruticans* (Gol pati), the feather palm or water coconut grows gregariously along the banks. *Phoenix paludosa* (Hatal) is seen dominant on drier border-lands.

The present work is based on information collected from tribal localities of Sajnakhali, Bakkhali etc. Earlier reports of Naskar and Guha Bakshi (1987) is remarkable. The tribal have their own system of medicines since the time immemorial.

2.1.2 An Ethno-Medico-Ecological Survey at Paschim Medinipur District For Natural Health Care and Green Belt Movement

Abstract

In the present paper the author has tried to unveil the potential value of different plants at Medinipur surroundings. While all most

Table 2.1.1

Ailments	*Local Name*	*Botanical Name & Family*	*Parts used*	*How medicine is prepared*	*How used*
Rheumatism	Dakur	*Cerbera manghas* (Apocynaceae)	Bark and milky latex	Crushed the bark and extract the juice	Externally smeared in the affected area
	Hargoza	*Acanthus ilicifolius* (Acanthaceae)	Leaf	Crushed the leaves and made into paste	-do-
Tiger-bite	Hargoza	-do-	Leaf	Extract the juice	Apply to the wound
Blood purifier	Hargoza	-do-	Fruit pulp		Eat daily till normal
Asthma	Dudi-lata	*Finlaysonia obovata* (Asclepiadaceae)	Leaf	Extract ½ cup juice	Fed to the patient once daily
	Kalo Baen	*Avicennia officinalis* (Avicenniaceae)	Leaf	Dry leaves are smoked.	Claimed for relief from asthma.
Leprosy	Kalo Baen		Leaf	Extract the juice	Fed to the patient once daily.
Milk production	Kalo Baen		Leaf	Fresh leaves are lactogenic both in Human & Cattle.	
Piscicide	Panilata	*Derris trifoliata* (Fabaceae)	Seeds	Crushed the seeds and used as fish poison	
Urinary problem	Khalamkuchi	*Boerhavia repans* (Nyctaginaceae)	Leaf	Crushed the leaves juice	-do-
Cold, Cough and diarrhoea	Sajina	*Moringa oleifera* (Moringaceae)	Leaf	Crushed the leaves juice	-do-

contd...

Table 2.1.1 – *contd...*

Ailments	*Local Name*	*Botanical Name & Family*	*Parts used*	*How medicine is prepared*	*How used*
Tonic	Alshi lata	*Mucuna gigantea* (Fabaceae)	Seeds	Extract ½ cup juice	Fed twice daily in protein deficiency
Tumour	Dhundul	*Xylocarpus granatum* (Meliaceae)	Seeds	Crushed the seeds and made into paste	Paste is used for relief of breast tumor
Groomig hairs	Dhundul	*Xylocarpus granatum* (Meliaceae)	Seeds	Oil extracted	Smear daily on the hair
Cataract	Champa	*Bruguiera parviflora* (Rhizophoraceae)	Fruit	Extract the Juice	Locally applied
Piles	Sundari	*Heritiera fomes* (Sterculiaceae)	Seeds	Crushed the seeds and made into paste	Locally applied
Ulcer, Haemorrhage	Goran	*Ceriops tagal* (Rhizophoraceae)	Bark	Decoction is used to cure malignant ulcers and stop haemorrhage.	
Snake-bite	Isharmul	*Aristolochia indica* (Aristolochiaceae)	Leaf	Crushed the leaves and made into paste	Smear on the wound as balm.

residents have knowledge of plants or plant parts use as medicine only local folk-doctors have the knowledge of season, place and mode of harvest, post harvest treatment, storage and uses doses. Commercial pressure has reached the area during the last decade and illegal export of a large number plant is reported. Suggestion for future management and a need of ethno-medico-ecological survey at Paschim Medinipur has also been discussed.

Keywords : Paschim Medinipur, Ethno-medicine, Biodiversity, Ecosystem, Ecological Balance.

Introduction

No attempt has yet been made (except Naskar, 1986, Ghosh, 2002, 2008) by any school and college students to discover and enlightened the Ethno-medicines of Paschim Medinipur district which are still hidden and popular among the tribals. Students of our eco-club are very much interested to survey these Ethno-medicines because which have no side effect to cure different ailments in human and animals in low cost and easy available in their surroundings. As the life saving drugs scarce and expensive, this system of treatment should be given incentive for further expansion.

Plants have been used as a source of medicine for living beings from ancient time. According to an estimate of the WHO, approximately 80% of the people in developing countries rely chiefly on traditional medicines for primary health care. Ethno-medico-ecological survey help mankind to search, develop new cures to ailment and to protect bio-diversity. As the traditional societies residing in micro-ecosystem have developed specific knowledge on plants and the present paper reports the information collected during this survey.

We all know trees abate pollution by serving as living filters of polluted air and can be called pollutant sinks. Species selected for green belt should be : -

a) Adopted to the local environment, Pollution tolerant and fast-growing.

b) Preferably broad-leaved as the large surface area of the foliage can absorb the gaseous pollutants.

c) Evergreen species with simple leaves having rough and fairy surface to collect dust throughout the year.

Recovery of the commons : Self-Help Movement to Rejuvenate our Biodiversity knowledge.

Does the epidemic of biopiracy imply we can do nothing ? No-self help and self determination is still the viable alternative as a strategy to fight biopiracy. The following are ways we can use the ethno-medicines in our homes and communities. Here is a opportunity for those who are keen to be knowledgeable.

Opportunity never wait for anybody, but it's we who have to grab it. Think how much poorer our nature as well as animals would be without them.

Material and Methods

Field survey was carried out between the years 2001-2003. Information on folk-medicinal uses of plants was obtained through interviews with local folk healers who often accompanied the author in the field. Data on the local name of the plant, parts used, other ingredients added (if any), method of preparation and mode of use etc. were recorded for each species. Samples of all folk drugs were collected with the help of informants.

Besides the present work is based on information collected from 56 students in their respective tribal localities of Keranichati, Gopegarh, Abas, Shekpura, Barua, Garhmal, Shiromoni, Karnagarh etc. The tribals have their own system of Ethno-medicine since the time immemorial. This survey aims to protect natures diversity, human & animal health and the environment through our consumption patterns (ensuring sustainable use of the natural resources and promoting sustainable life styles).

Results and Discussion

39 plant species were recorded for various ailments. Data on the medicinal uses of the plants are presented in table.

The villages of the region are rich in ethno-medicine knowledge owing to their close affinity with the surrounding plant cover. They obtain a variety of plant products from wild to fulfill their own needs as they are economically weaker sections of the society. In the tribal areas the rules and regulations by which the tribal people have been traditionally governed are now being gradually abolished by the young literate generations. Another crucial factor responsible for such change is the migration of youth from tribal areas to urban centres to take up jobs. This gap is further accentuated by the adoption of modern trend. Therefore, the importance of recording indigenous knowledge base related technology as described here becomes essential in view of rapid socio-economic and cultural changes and for high tech low cost solution. Religious - cultural faith, weak economy and consequently lack of modern medical opportunities in these villages seem to be the cause of depending on these plants. While conducting the survey, reveals that most of the people were found dependent on plants and they also preferred it, although the preparing methods are known only to local faith - healers. Due to rapid pollution, population, civilization, industrialism and excessive exploitation of plants-animals by the poachers and other plant-animal species are now limited in their natural habitats. Some species have become rare, threatened and endangered. Such as *Drosera burmannii, Trapa bispinosa,* Bagrol, Pangolin, etc.

This survey movement is committed to :-

Conservation of biodiversity and re-introduction of native crop and plant varieties on our farms, lands and forests. Encouraging and supporting small farmers, tribals and folk-doctors to cultivate the indigenous medicinal plants to save their folk-medicinal practices. Our Eco-club is dedicated to sustainable rural development and also to bring social, economical, environmental awareness among us.

According to an on going study of the Ministry of Environment and Forests, Govt. of India, 17000 species of higher plants in India 7,500 species have been recorded to have medicinal value. These medicinal plants grow in a variety of eco-systems in various Bio-geographic Zones.

Conclusion

This survey looks to be a significant to enriching the economy. Medicinal plant cultivation under land rehabilitation and National Green Crops programmes through people - student's participation is an encouraging tool for the improvement of the economic status of local inhabitants, biodiversity and ecological balance. The medicinal plant cultivation have higher monetary output/input ratio than those of each crops and horticultural crops.

Table 2.1.2

Ailments (Medical efficacy claimed)	*Local Name*	*Botanical Name & Family*	*Parts used*	*How medicine is prepared*	*How used (Mode of use)*
Malignant tumour	Sasha	*Cucumis sativus* (Cucurbitaceae)	Fruit	One fresh fruit (100 gm)	Ate once daily to prevent from cancer due to presence of Cucurbitasin
Malignant tumour	Bilati Begun	*Lycopersicon esculentum* (Solanaceae)	Fruit	One fresh fruit (100 gm)	Claimed as a anticancer due to presence of of licopen
Snakebite (Boas)	Kantal	*Artocarpus heterophyllus* (Moraceae)	Peduncle	Extract the juice	Fed 1 cup juice thrice daily immediately after bite
Snakebite	Iswarmul	*Aristolochia indica* (Aristolochiaceae)	Bark, root	Extract the juice	Fed to the human and cows ½ cup juice twice daily till cure
Food poison	Arimed	*Acacia leucophloea* (Leguminosae)	Bark of latex	Extract the juice	Fed it immediately
Blood vomiting	Berela	*Sida cordifolia* (Malvaceae)	Root, leaf	Crushed and made into paste	Fed it immediately
Accumulation of fat	Nishinda	*Vitex negundo* (Verbenaceae)	Leaf	Extract the juice	One teaspoonful fed daily
Accumulation of fat	Mangustan	*Garcinia mangostana* (Guttiferae)	Leaf, husk	Extract the juice	One teaspoonful fed daily

contd...

Table 2.1.2 – *contd...*

Ailments (Medical efficacy claimed)	*Local Name*	*Botanical Name & Family*	*Parts used*	*How medicine is prepared*	*How used (Mode of use)*
Pain in teeth (toothache)	Gorap – Begun	*Solanum surattense* (Solanaceae)	Root	Collect the fresh root	Grind the root by the affected tooth
Anti–fertility	Nagdona	*Artemisia vulgaris* (Compositae)	Leaf, root	Crushed and made into tablet	10 tablets in each cycle claimed to be contraceptive
Impotence	Shialkanta	*Argemone mexicana* (Papaveraceae)	Seed, latex	Crushed the seed with its latex	5 gm paste taken daily for 30 days to cure
Blood dysentery/ Diarrhoea/Night wetting	Ramdantan	*Smilax indica* (Liliaceae)	Root	Extract the juice	Drank ½ cup juice twice daily till cure
	Ulatkambal	*Abroma augusta* (Sterculiaceae)	Root	-do-	-do-
Wasp sting	Baichi	*Flacourtia indica* (Flacourtiaceae)	Stem bark	Fresh bark	Grind the bark by teeth slowly
Alopecia/Boils/ Wounds	Karanj	*Caesalpinia crista* (Caesalpiniaceae)	Seed	Oil is extracted from the seed	Smear the oil in hair, boils and wounds till cure
Blood sugar	Karanj	*Caesalpinia crista* (Caesalpiniaceae)	Leaf	Crushed 10 leaves and made into paste	Fed to the patient once daily for 7 days
Dandruff/ Premature hair greying	Sikaki	*Acacia concinna* (Leguminosae)	Leaf, seed	Extract the juice	Regular external applications are hair vitalizer

Table 2.1.2 – *contd...*

Ailments (Medical efficacy claimed)	*Local Name*	*Botanical Name & Family*	*Parts used*	*How medicine is prepared*	*How used (Mode of use)*
Miscarriage	Anantamul	*Hemidesmus indicus* (Asclepiadaceae)	Root	2 ½ gm roots of anantamul and 2 gm bark of daruchini boiled in a glass of cow's milk	Fed to the patient once a day for 7 days
	Daruchini	*Cinnamomum zeylanica* (Lauraceae)	Bark		
Acne	Ayapan	*Eupatorium ayapana* (Compositae)	Leaf	Extract the juice	Externally applied to the face
Alopecia	Datura	*Datura metel* (Solanaceae)	Leaf	Extract the juice	Smear the juice in head for 30 minutes
Acne / Alopecia	White Sarisha	*Brassia campestris* (Cruciferae)	Seed	Crushed the both seeds (1:1) and made into paste	Externally used in head and face
	Til	*Sesamum indicum* (Pedaliaceae)	Seed		
Miliaria rubra	Anantamul	*Henidesmus indicus* (Asclepiadaceae)	Root	Extract the juice	Smear on the body
Kidney stone	Mashkalai	*Phaseolus mungo* (Papilionaceae)	Cotyledon	Cotyledons soak-ed for overnight	Drunk the leached water at dawn
Food poison	Iswarmul	*Aristolochia indica* (Aristolochiaceae)	Bark, root	Extract the juice	Fed to the cows
Diarrhoea	Swarnạlata	*Cuscuta reflexa* (Cuscutaceae)	Stem	Extract the juice	Fed to the cows thrice daily till cure

contd...

Table 2.1.2 – *contd...*

Ailments (Medical efficacy claimed)	*Local Name*	*Botanical Name & Family*	*Parts used*	*How medicine is prepared*	*How used (Mode of use)*
Infertility	Bon – Dhenros	*Malachra capitata* (Malvaceae)	Fruit	Collect fresh fruit	Fed to the patient 5 raw fruits daily during mens period for 3 months
Asthma	Dayalu flower	*Acrua lanata* (Amaranthaceae)	Flower with leaf	Extract the juice	Drunk the juice with few drops of honey for one month
Infertility	Swet – Lajjabati	*Mimosa pudica* (Mimosaceae)	Root	Extract the juice	Drunk one teaspoonful juice alongwith a pepper for 20 days
	Swet – Aparajiata	*Clitoria ternatea* (Caesalpiniaceae)	Root	Extract the juice	- do -
Low pressure	Bon – Kalmi	*Ipomoea paniculata* (Convolvulaceae)	Leaf	Extract the juice	Drunk ½ cup juice once daily for 15 days
Enlargement of liver / Dim sighted	Boncharal	*Desmodium gyrans* (Desmodiaceae)	Leaf	Extract the juice	Drunk 2 spoonful juice daily
Pox	Swet – Kantikari	*Solanum xanthocarpum* (Solanaceae)	Arial part	Crushed and made into paste	2 dry tablets will be eaten for 7 days as an antidote both in human and cow
Hydrophobia	Bans	*Bambusa vulgaris* (Gramineae)	Root	Crushed in 1:1 ratio	Orally used and externally applied to the wound as an antidote both in human and cow

contd...

Table 2.1.2 – *contd...*

Ailments (Medical efficacy claimed)	*Local Name*	*Botanical Name & Family*	*Parts used*	*How medicine is prepared*	*How used (Mode of use)*
	Ankar	*Alangium salviifolium* (Alangiaceae)	Root		
Snake – bite	Barachadar	*Rauwolfia canescens* (Apocynaceae)	Root	Extract the juice	Drunk the juice and also smear the juice in the wound
Blood sugar	Barachadar	*Rauwolfia canescens* (Apocynaceae)	Root	Extract the juice	Fed to patient alongwith *Terminalia arjuna* bark
Diabetes	Gudmar	*Gymnema sylvestre* (Asclepiadaceae)	Leaf, fruit	Extract the juice	Drunk the juice
Constipation	Golmarich	*Piper nigrum* (Piperaceae)	Fruit	Crushed 3–4 fruits	Drunk the dust during bed time at night with a cup of lukewarm water
Flatulence	Kul	*Ziziphus jujuba* (Rhamnaceae)	Leaf	Made into paste	Rubbed the frothy paste in abdomen
Toothache	Peypey	*Carica papaya* (Caricaceae)	Latex	Collect fresh latex	Add a drop of latex on infected teeth

2.1.3 A Contribution to the Ethnobotany of Gujarat: Some Medicinal Plants Used as Antidiabetic, Antipyretic and Icteric Agents

Abstract

The present paper places some valuable information of 28 plant species (belonging to 27 genera under 23 families) of Gujarat having ethno-medico-botanical value. Seven plants are used as anti-diabetic, seventeen as antipyretic and nine as icteric agents. Plants are listed alphabetically followed by vernacular names (in Gujarat), Botanical families and their uses.

Key words: Medicinal plants, Ethnobotany, Gujarat.

Introduction

In the last few decades ethno-botanical studies have received immense attention to bring back the centuries' old heritable knowledge on plants, especially from tribal communities who live in remote villages and have intimate relation with the plants around them (Faulks, 1958; De, 1968; Beatrice, 1975; Jain 1981; Anonymous 1983; Satyabati, 1991; Husain, 1992; McIntyre, 1994; Bhattacharya, 1996; Cotton, 1996; Huang, 1999; Bhattacharya, 2002 and many more). Several most potent remedies used today, e.g. aspirine, atropine, cocaine, codeine, digoxin, emetine, ephedrine, morphine, pilocarpine, vinblastine, vincristine etc. are derived from plants (Sukhdev, 1997). The present investigation is aimed to create awareness about the ethno-medico-botanical value of 28 plants from Gujarat and to draw the attention of phytochemists and pharmacologists.

Land and People

Gujarat is situated between 24°36' N and 20° 02' N latitude 38° 08' E and 74° 24' E longitude. This state is divided into three geographical regions. (i) Gujarat mainland, mostly alluvial terrain situated between Narmada and Tapi rivers and bounded by the hills of Aravalli, Satpura, Vindhya and Sahyadri. (ii) Small peninsular Kachchh on the North - West with the famous "Rann of Kachchh" and (iii) Saurashtra peninsula, a semiarid ecoclimatic region central part of which is hilly and undulating land surrounded by sea from three sides. The main underlying rock system is mostly basalt. The typical clayey black cotton soil is derived from this rock and distributed all over except Kachchh and the coastal areas, where it is sandy and alkaline alluvial. The rainfall is most irregular, highest in South Gujarat and the lowest in Kachchh. Except in the arid regions of the north and northwest rainfall varies between 650 mm and 1270 mm. The maximum temperatures are recorded lower in southern parts and higher up to 44°C in northern parts.

Gujarat has considerable large tribal population, about 14.2% of the total population of the state (Anonymous, 1986) e.g. Dholia, Dubla, Gamit, Koli, Mer, Nayakad, Rabari, Rathawa, Siddi and Varli (Naik, 1951 and Sinha, 1976). They do possess considerable ethnobotanical knowledge including traditional practices, most of these informations remain undocumented (Abdi, 1993). The lacunae in information concerning medicinal plants of the state necessitated undertaking the present investigation.

Methods

This information has been collected from many people of various communities and consulting the "Native Doctors" such as 'Bhagat', 'Bhua', 'Vaidya', as they possess inherited knowledge regarding the plants of Ethnobotanical importance. However, the following claims deserve further detail studies since a proper scrutiny only can decide the therapeutic potentiality. The list of all the species presented in this paper was prepared after full identification with the help of the literature. (Cooke, 1958; Santapau & Janardhanan, 1967; Shah, 1978; and Bole & Pathak, 1988) and the help of the taxonomists of the Western Regional Circle, BSI, Pune was also taken for identification.

Enumeration

The botanical name of the plants are arranged in alphabetical order followed by local name, name of the family and the part(s) used (cf. Table 2.1.3). People of this region, especially tribes, have developed their traditional practices since time immemorial. Use of certain plants for some particular medicines is restricted to some people only and normally they are quite averse to share their knowledge with any outsider. Therefore, there is an urgent need of documentation of this irreplaceable knowledge. Day after day with the advent of modernization it may be lost forever.

Table 2.1.3 : List of the Plants of Ethno-Medico-Botanical importance

Sr. No.	*Botanical Name*	*Vernacular Name*	*Botanical Family*	*Part(s) used*
ANTIDIABETIC				
1.	*Azadirachta indica* A. Juss	Kadvo Limbo	Meliaceae	Stem bark extract
2.	*Enicostema hyssopifolium* (Wild) Verdoon	Memejevo, Kadvi Nai	Gentianance	Decoction of leaves
3.	*Ludwigia perennis* L.	Panlavang	Onagraceae	Whole plant
4.	*Mimosa pudica* L.	Lajamani	Mimosaceae	Dried root powder
5.	*Maytenus emarginata* (Wild) D. Hou	Vicklo	Fabaceae	Water kept in a cup made up of heart wood and "Gum kino"
6.	*Pterocarpus marsupium* Roxb	Biyo Piasal	Fabaceae	Seeds and tender leaves
7.	*Trigonella foenum graecum* L.	Methi	Fabaceae	Seeds and tender leaves
ANTIPYRETIC				
1.	*Adansonia digitata* L.	Rukhdo	Bombacaceae	Root decoction
2.	*Ailanthus excelsa* Roxbi	Moto Arduso	Simaroubaceae	Leaf decoction
3.	*Averrhoa carambola* L.	Kamarak, Kamranga	Oxalidaceae	Fruits
4.	*Azadirachta indica* A. Juss	Kadvo Limbdo	Meliaceae	Stem bark
5.	*Boerhavia diffusa* L.	Satodi	Nyctaginaceae	Root powder
6.	*Biophyturn sensitivum* (L) DC.	Risamnu, Zarero	Oxalidaceae	Leaf decoction
7.	*Centella asiatica* (L) Urb.	Brahmi	Apiaceae	Juice of whole plant
8.	*Clematis hedysarifolia* DC.	Morvel	Ranunculaceae	Decoction of whole plant
9.	*Eclipta alba* (L) Hassk.	Bhangro	Asteraceae	Juice of fresh leaves warm by hot iron object
10.	*Enicostema hyssopifolium* (Wild) *Verdoon*	Mamejevo, Kadvi Nai	Gentianaceae	Decoction of leaves

contd...

Table 2.1.3 – *contd...*

Sr. No.	*Botanical Name*	*Vernacular Name*	*Botanical Family*	*Part(s) used*
11.	*Holarrhena antidysenterica* (Heyne ec Roth) wall	Indrajav	Apocynaceae	Seeds
12.	*Hybanthus enneaspermus* (L) F. Muell	Banafsha	Violaceae	Decoction of leaves
13.	*Leucas aspera* (wild) Spr.	Kubi	Lamiaceae	Decoction of leaves
14.	*L. urticaefolia* R.Br.	Kubo	Lamiaceae	Decoction of leaves
15.	*Ocimum canum* Sims	Ran Tulsi, Jangli Tulsi	Lamiaceae	Decoction of leaves
16.	*Semecarpus anacardium* L.F. Suppl.	Bhilamo	Anacardiaceae	Fruits
17.	*Tinospora cordifolia* (Wild) Miers ex HK.	Galo	Menispermaceae	Stem decoction
ICTERIC				
1.	*Alangium salviifolium* (L.F.) Wang	Ankoli	Alangiaceae	Root bark
2.	*Cassia fistula* L.	Garmalo	Caesalpiniaceae	Sticky black pulp of legume and jaggery
3.	*Curculigo orchioides* Gaertn	Kali Musli	Hypoxidaceae	Underground bulb
4.	*Eclipta alba* (L) Hassk.	Bhangro	Asteraceae	Juice of fresh leaves
5.	*Leucas urticaefolia* R. Br.	Kubo	Lamiaceae	Decoction of leaves
6.	*Maytenus emarginata* (Wild) D.C.	Vicklo, Vickro	Celastraceae	Leaves are chewed
7.	*Oxystelma secamone* (L) Karst.	Lal Dudhi	Asclepiadaceae	Fresh root
8.	*Rhynchosia minima* (L) DC	Nani Kamalvel	Fabaceae	Fresh root
9.	*Trianthema portulacastrum* L.	Satodo Sapunne	Aizoaceae	Juice of fresh plant

2.1.4 Fruit Can Play as an Antidote in Different Ailments

Different fruits used by the tribals and local communities of Bankura and Paschim Medinipur districts, West Bengal, are antidotes have been reported.

Key Words : Bankura, Paschim Medinipur, Lok Vaidyas.

Introduction

Fruits are certainly a useful indicator. The present paper is an attempt to enlist such activities with a view to provide further investigations in the search of new drugs to combat these disorders. Some earlier reports are also available (Billore 1980, Gillam 1980, Sen and Batra 1997, 1998).

Methodology

Fieldwork was carried out between the years 2000-2003. During the field trips interviews were taken with folk doctors, chieftains, old experienced persons for documenting home remedies.

Enumeration / Local Therapy

The plants are arranged by their botanical names followed by their local names, family and the part(s) used. Use of certain fruits for some particular medicines is restricted to some people only and normally they are quite averse to share their knowledge with any outsider. Day after day with the advent of civilization it may be lost forever. This therapy is cheap, readily available on their doorsteps and moreover, they have traditional belief on this treatment.

The healers suggest to use luke-warm water for drinking and the avoidance of oils and chillis. If a patient doesn't get relief from ailment by a particular drug in specified period of time, then another drug is tried by the healers.

Discussion

The ethnic communities have the impression that the diseases are caused due to curse of gods, goddesses and spirits. So, they first seek magical and religious practices through the "Lok Vaidyas" to get rid of ailments. These "Lok Vaidyas" are the faith healers in one hand and herbal doctors on the other hand. The encroachment of forest lands for expansion of industry, agriculture and

Table 2.1.4

Local name of fruits	Botanical name and family	Medical efficacy claimed	Mode of use
Kagzi Lebu	*Citrus aurantifolia* (Rutaceae)	Bulky body, high blood pressure	Mixed one teaspoonful honey and 4 teaspoonful lime juice in a glass of lukewarm water and drunk it daily at dawn in empty stomach.
Batabi lebu	*Citrus maxima* (Rutaceae)	Jaundice	Juice is very effective for the patient.
Bel	*Aegle marmelos* (Rutaceae)	Dysentery	Mixed the dust of bel and isabgul (1: 1) in a glass of lukewarm water and drunk it during bed time at night.
Kanthal	*Artocarpus heterophyllus* (Moraceae)	Weakness	Drunk ½ cup juice with a cup of lukewarm cow's milk.
Dalim	*Punica granatum* (Puniaceae)	Insomnia, Narcotic, Diarrhoea, Dysentery	Drunk ½ cup juice along with extract of Ghrita Kumari (Aloe vera). Ate rind of the fruit with clove.
Kala	*Musa sapientum* (Musaceae)	Kidney, Liver	Daily intake of a fruit claimed to keep healthy both liver and kidney.
Kanchkala	*Musa paradisiaca* (Musaceae)	Anaemia	Fed it daily through soup or boiled rice.
Peypey	*Carica papaya* (Caricaceae)	Dyspepsia, Liver disorder	Boiled raw fruits claimed to cure dyspepsia. Latex of fruits (4-5 drops) with sugar-cake.
Sasha	*Cucumia sativus* (Cucurbitaceae)	Blood sugar and Cholesterol	Ate once daily without salt.
Kamla lebu	*Citrus reticulata* (Rutaceae)	Mental depression, Constipation	Boiled the rind in water and then bath it in cool condition. Boiled the fruit in water and then drunk it.
Jam	*Syzygium cumini* (Myrtaceae)	Blood sugar, Dysentery	Boiled 1 gm seed dust alongwith 5 ladies fingers and fed it daily.
Dhenros	*Abelmochus esculentus* (Malvaceae)		

contd...

Table 2.1.4 – *cont...*

Local name of fruits	*Botanical name and family*	*Medical efficacy claimed*	*Mode of use*
Apple	*Malus sylvestris* (Rosaceae)	Constipation	Ate 1 apple daily along with the rind
Amlaki	*Emblica officinalis* (Euphorbiaceae)	Diabetes	5 gm amlaki ate daily alongwith pinch of lime and honey
Akhrot	*Juglans regia* (Juglandaceae)	Worm infection	Decoction of fruits
Tut	*Morus alba* (Moraceae)	Laxative, Biliousness sensation	Drunk 1 cup juice thrice daily
Safeda	*Achras zapota* (Sapotaceae)	Blood dysentery	Drunk 1 cup juice thrice daily
Baghnokh	*Martynia diandra* (Pedaliacea)	Snakebite	Fruit will be chewed
Kala	*Musa sapientum* (Musaceae)	Contraceptive	Watery substance of the pseudostem (2 ml for 7 days)

rehabilitation have marked impact in depletion of fruit yielding plants. The smuggling of forest products, excavation of stones, gravel and red soil etc. are also the factors.

It is compulsory to get license to collect herbal drugs from the natural resources. For this some royalty is levied for each drug. The execution of these rules depends upon the awareness of the local people. Considering the depleting forest and the plant resources, efforts are being initiated by different agencies to protect plant diversity.

2.1.5 Instant Household Remedies for Dyspepsia and Cooking Industry

Household remedies from different food and fruits by the local communities of West Bengal against dyspepsia and every day needs have been reported.

Key Words: Senescence, Household remedy, Dyspepsia.

Introduction

Food have been used as a source of medicine by human from the time immemorial. The indigenous knowledge of traditional communities were formulated (Prakash -1998). According to WHO approximately 80% of the people in developing countries depend on traditional knowledge i.e. instant household remedies for primary health care (Anonymous -1998). These household remedies have developed specific knowledge. In our daily diet we ate different food and drinks. But all these foods and drinks are not always suitable for all due to different health conditions. Traditional knowledge are very scientific and effective in our every day life.

Methods

These informations have been collected from many people of local communities and by consulting the folk-doctors.

However, following claims deserve after proper investigation among different age groups.

Results

Data for instant household remedies are presented in table 2.1.5 a & b along with mode of use. The observation undertaken during the period of 2000-2003.

Discussion

Information mentioned herein is based on personal observation and from local healers who have long been using this technique which happens to be the most common difficulties and health problem among different ages of man in their daily life. Here all the drugs are administered orally and methods of drug preparation are very simple. Drugs are prepared either by grinding, crushing, extraction or by decoction. Such investigations may bring to light new natural method with potential for wider application for dyspepsia and for our daily needs in house.

2.1.6 Potential Remedies for Everyday

Abstract

Instant household remedies have been reported which are very essential in daily life for housewives.

Keywords : Household remedy, edible oil, icing, rice gruel, sacrificial knife.

Table 2.1.5 a : Remedies for Dyspepsia

Types of food intake	*Quick remedies*		*Mode of use as orally administered*
	Local Name	*Botanical name and family*	
Flour	Sasha	*Cucumis sativus* (Cucurbitaceae)	Fresh fruit
Rice	Joan	*Trachyospermum amni* (Umbelliferae)	Joan alongwith a pinch of salt
Herbage vegetables (Sag)	Sarisha	*Brassica campestris* (Cruciferae)	Seed paste
Fish, meat	Stale rice water		
Oil	Pan	*Piper betle* (Piperaceae)	Leaf
Sweet	Labanga	*Syzygium aromaticum* (Myrtaceae)	Clove alongwith water
Kul (Jujube)	Hot water		Drink the luke warm water
Cake	Hot water		Drink the luke warm water
Ghee	Gora lebu	*Citrus acida* (Rutaceae)	
Milk	Butter milk		
Banana	Rock salt		
Mango	Hot milk		
Coconut	Boiled rice		
Jam (Black plum)	Haritaki	*Terminalia chebula* (Combretaceae)	
Matar (Pea)	Haritaki	*Terminalia chebula* (Combretaceae)	
Kanthal (Jack fruit)	Kala	*Musa sapientum* (Musaceae)	
Ol (Elephant foot)	Molasses (akh)	*Saccharum officinarum* (Gramineae)	
Afing, Opium,	Milk / Tentul	*Tamarindus*	

contd...

Table 2.1.5 a – *contd...*

Types of food intake	*Quick remedies*		*Mode of use as orally administered*
	Local Name	*Botanical name and family*	
Poppy		*indica* (Caesalpiniaceae)	
Pulse	Joan	*Trachyospermum amni* (Umbelliferae)	Joan alongwith a pinch of salt
Rice milk	Uncooked sonamug dal	*Phaseolus aureus* (Papilionaceae)	
Pungent curry	Mustard oil		
Hotch potch	Rock salt		

Table 2.1.5 b : Potential remedies for daily needs in kitchen

Sl. No.	*Problems*	*Household remedy*
1.	Maltation of salt due to humid climates of rainy season	1. A shred of cloth containing rice placed in the containing salt-container
2.	Quick senescence of ginger *(Zingiber officinale)*	2. Buried the gingers within the sand for its freshness
3.	Hasten senescence of chillies *(Capsicum frutescens)*	3. Detached the petioles from the fruits
4.	Hasten senescence of Kanch-kala *(Musa paradisiaca)* and Kagzi Lebu *(Citrus aurantifolia)*	4. To keep freshness immersed the fruits daily in cold water for an hour
5.	Promote senescence of bitter gourd *(Momordica charantia)*	5. Cut the fruits in two equal halves longitudinally and deseeded
6.	Decaying of eggs	6. Smear the gum on outside the egg or immersed the eggs for few days in saline water or limy water.
7.	Delay in lightening of coal-wood woven	7. Spray some amount of salt in the woven
8.	Sear in the utensil	8. Rubbed the utensil with wheat bran

contd...

Table 2.1.5 b – *contd...*

Sl. No.	*Problems*	*Household remedy*
9.	Cracking in glass due to hot liquid	9. Placed a metal spoon in the glass before filling with liquid
10.	Get flabby of the match-stick during rainy season	10. Keep a few sunned-rice within the match-box and then wrapped the box with a shred of cloth
11.	Get flabby of the biscuits	11. Keep the biscuits in casket alongwith some amount of sugars or blotting papers. Keep the biscuits within a fried-rice container to keep the biscuits in crisp condition
12.	Spoiling of excessive ground spices for cooking	12. Stored the ground spices for cooking with a small amount of mustard oil or salt.
13.	Turmeric *(Curcuma longa)* smell in the curry	13. Immersed a burnt spud in the hot curry or covered a tender banana leaf on the curry
14.	Be filled with tears due to peel off the rind of onion	14. Cut the onion stem into 2 equal halves and then immersed in water for 5 minutes
15.	Stored grain pest occurs in rice, wheat, pulse	15. Within the grains sac keep some amount of dry neem leaves or dry chillies
16.	Quick burning of wax-candle	16. Before set on fire of the wax-candle sink it in water for its longevity
17.	To prepare cheese	17. Add fresh latex of *Carica papaya* in milk

Introduction

The indigenous knowledge of traditional communities were formulated by Prakash, 1998. According to W.H.O. approximately 80% of the peoples in developing countries depend on traditional knowledge for primary health care (Chattopadhyay and Maji 1975, Anonymous, 1998). Due to excessive cost of the medicine traditional knowledge are very popular and effective for its cheapness and easily available in rural areas.

Methods

These informations have been gathered from old-aged experienced housewives after proper investigation. The arguments based upon coincidence of facts. In the remote villages these knowledges are still effective as first-aid.

2.1.7 How You Can Survive for a Long Time in the Earth?

Introduction

Ageing refers to the sequential changes that occur in a plant or animal as it proceeds from zygote formation to death. Whereas senescence is a phase of the ageing process which is clearly degenerative - ultimately irreversible.

Table 2.1.6 : Potential remedies for everyday life

Sl. No.	*Problems*	*Household remedy*
1.	Pest control for cloth	- Keep the dry rind of orange with the cloth.
2.	To get more soury juice from tamarind.	- Immersed the tamarind for 5 minutes in cold water and then boiled.
3.	Any type of tea-scar on shirt or cloth.	- Rubbed the scar with a few drops of glycerine. After 5-6 hrs wash it by the detergents or soap.
4.	To keep the room free from ants.	- Spray the room with tobacco leached water.
5.	To keep the boiled egg in edible condition for 3-4 days which is often essential for hotel, picnic, feasts, etc.	- The intact boiled eggs immersed in cold water and keep into the refrigerator. In lack of refrigerator smear mustard oil or dalda throughout the outer surface of the egg.
6.	Hasten cooking of the Arhar pulse *(Cajanas cajan)*	- Boiled or cooked the pulses in rice leached water.
7.	To keep the kitchen and dining room free from house-fly.	- Spray the room with few drops of freshly crushed neem leaves (*Azadirachta indica*) water.
8.	To keep the edible salt free from clot.	- Fry the salt and then keep it in a container.

contd...

Table 2.1.6 – *contd...*

Sl. No.	*Problems*	*Household remedy*
9.	To increase the merits of tea taste	- Add a pinch of ripe sour wood apple *Feronia elephantum* (Kathbel) after tea-making.
10.	To keep the excessive cut Kagzi lebu in good condition in feast or hotel	- Keep the cut pieces of lime lemon *(Citrus aurantifolia)* within a edible salt container.
11.	To keep the kitchen room free from house-lizard.	- Place a piece of cock-egg shell in their entrance and in ventilator.
12.	Pest control in cloth	- After final washing of cloth through detergent or soap immersed the cloth in a basket of water with 5 drops of kerosene.
13.	To save the consumption of edible oil along with free from excessive fat intake.	- Before cooking of cashew nut or groundnut smear the nuts with few drops of edible oil and then roast it without oil.
14.	Proper utilization of stale loaf or bread.	- Keep the stale breads in a steel - strainer and apply hot vapour to it to regain its spongy and good condition. Dry breads crushed into dusts and use it in chop, cutlet, etc.
15.	Preparation of cake without egg	- Use milk powder and corn flour in 1:2 ratio.
16.	To keep the excessive uncooked egg-yolk in good condition.	- Immersed the yolk in cold-water for few days in lack of refrigerator.
17.	Burning sensation of hands during the preparation of chillies and peppers.	- It can be cured if the hand is immersed in cold milk for few minutes.
18.	For easy peel off the rind of garlic in hotel and feast.	- Immersed the whole garlics in boiled water for 2-3 minutes and then rubbed the rind to crush it.
19.	To keep the salad free from bacteria	- Immersed the salads for few minutes in cold water containing vinegar before intake.

contd...

Table 2.1.6 – *contd...*

Sl. No.	*Problems*	*Household remedy*
20.	To get onion smell and taste in lack of onion	- Immersed the hing (*Ferula asafoetida*) in ginger's juice and then use it in curry.
21.	Curd for blood sugar and diabetic patient, to make the curd fresh and sweety.	- Hung the curd by cloth for half an hour. Add milk in curd according to the requirements.
22.	To increase the longevity of sauce.	- During the preparation of sauce add a pinch of vinegar in oil.
23.	To get rid of mould use cold oil.	- During the preparation of sauce boiled the oil and then cool it.
24.	To get rid of from swell up or boil over the milk.	- Keep some amount of butter inside the upper eave of the container during the boiling of milk.
25.	Keep aloof from or keep away from pest.	- Mixed a pinch of marine salt with the rice.
26.	For the preparation of quick icing.	- Smear the honey alongwith the ingredients of groundnuts to make the cake tasty.
27.	To increase the longevity of ghee.	- Add a piece of molasses in ghee to keep it fresh for long time.
28.	To get rid of from often swell-up or boil over the rice gruel.	- Rubbed the outer surface of the container/cooking pot with a piece of wet cloth.
29.	To get rid of from the hardy problem of the globules of the pulses and cheese within curry.	- At first made the soup according to the requirements and then placed the fried globules in the soup in warm condition.
30.	Decaying of coriander leaves.	- Immersed the leaves in cold water and then dry in air. Detached the petioles from the leaves and put it into plastic bag.
31.	Excessive burning scar of the pan.	- It can be prevented if the empty pan heated at first deeply and then filled up the pan with oil according to the requirement for cook.

contd...

Table 2.1.6 – *contd...*

Sl. No.	*Problems*	*Household remedy*
32.	Insect bite.	- Smear a pinch of tobacco paste in the wound.
33.	Pest control for clothes.	- Insert dry rind of goralebu (*Citrus limon*) among the cloth.
34.	In piles.	- After sacrification of a goat in Kali-temple immediately wash and drink water of sacrificial knife.
35.	In dysentery.	- Drink the juice of crushed leaves of Ayapan.
36.	To keep your pulses and coarse flour of wheat free from pest.	- Roasted slightly and put it in a container.
37.	In cut and wounds of limbs.	- Latex of lonkasij (*Euphorbia tirucalli*, Euphorbiaceae) used externally.
38.	In ear pain.	- Give a drop of fresh woman's milk.
39.	In low hearing.	- Smear the paste of black cumin and seeds of bitter gourd in the base of external ear.
40.	In snake-bite.	- Fed the patient tobacco-ashes with water.
41.	In dog-bite.	- a) Fed the patient thorn apple's root with milk. b) Smear the paste of garlic in the wound. c) Leaf juice of *Aloe indica* alongwith rock-salt smear on the wound as balm for the 3 days.
42.	For easy delivery.	- a) Wear the root of banana (having seeds) in neck. b) Wear the root of *Borussus flabellifer* in waist which is situated in north side of dwelling house.
43.	Oozes out of milk from mammary-gland.	- Smear the paste of *Ricinus communis* (castor) leaves.

contd...

Table 2.1.6 – *contd...*

Sl. No.	*Problems*	*Household remedy*
44.	To keep the house free from snake and scorpion.	- Hung the horns of twelve horned deer.
45.	For easy digestion of jackfruit.	- Before intake immersed the deseeded seed vessels in boiled milk for 10 minutes.
46.	To get rid of from $CuSO_4$.	- Immersed the ripe mango in cold water for 15 minutes before intake.
47.	Keep under control the goats from *Ipomoea fistulosa* leaves.	- Leaves caused death of the goat in feeding.
48.	Keep aloof from shakhalu seeds (Yam Bean)	- Seeds of *Pachyrhizus angulatus* caused death of the human after intake.
49.	Remedy for locked jaw.	- Add a pinch of marine salt with a finger.

In normal condition the genetically programmed deteriorative processes involving loss of self-maintenance that ultimately lead to death is known as senescence. From thermodynamic standpoint increase in entropy is associated with the process. Among the many facets of physio-chemical events nothing is more interesting than the study of senescence as it not only tells us the mystery of death of an organism but also gives an understanding about the different aspect of correlative plant and animal behaviour which maintain biological diversity. Research to increase the longevity has been a pet topic. As a consequence 50 years ago a discipline called gerontology (the science of ageing) has been evolved.

Human from time immemorial struggling to increase their individual existence in the earth though death is inescapable. Now gerontologists all over the world so much busy to unveil the secrets of ageing as well as took efforts to stall or at least to retard the process of ageing (senescence). According to Dr. Richard Cutler of the National Institute of Ageing (NIA), Gerontology Research Center, Bethesda, USA, the goal of gerontologists is not to simplify to increase longevity but to ensure disease free old man. Better medical health cares, nature care (naturopathy), acupressure, natural food and healthy pure drink, tension free, calm and healthier i.e. unstressed healthier life styles, pollution free natural environment, mild physical exercise and walking, vegetarian with under diet (over eating causes premature ageing), the men who are very close to nature, periodical or intermittent fasting are always beneficial. Semi starved and often fasting, a restricted diet but nutritionally adequate somewhat deficient in calories particularly during the early ages lengthened the life span, make healthier, fertile, strong in immunity and delayed ageing. The above processes certainly contributed to increase for adult longevity though gerontologists claimed that human body is controlled by biological clock which is genetically programmed for each family to cease life despite's man's best efforts to intervene. Scientists have proved that death though inevitable, isn't predetermined. Gerontologists anticipate human longevity could be extended to 100 years or even more. They have identified so many causes which seems to culminate the ageing process (WHO 1980). Restriction of food intake delayed the rate of ageing and age-related diseases alongwith physiological deterioration of various organs (kidneys, lungs, heart, eyes, ears, etc.) and systems (alimentary, excretory, circulatory, etc.).

High level of antioxidants retard senescence because it attacks free radicals, the poisonous byproducts of biochemical reactions such as breathing, nutrition, circulation, excretion and other physiological activities. Free radicals are the prime cell killers in connection with or pertaining to ageing, senescence, disease and longevity. The dietary antioxidants stop or neutralize the activity of oxygen free radicals (which is formed during ATP formation as byproduct) thereby increase the longevity and to prevent old age degenerative changes. Antioxidants (butylated hydroxyanisolein edible oils) are also used to retard rancidity in fat and oils and minimize the oxidative destruction of vitamins and essential fatty acids. Human growth hormones from pituitary gland and sex hormones influence ageing by increasing the lean body mass, decreased body fat and thereby show a more youthful appearance, generally reproductive organs directly dictate the onset of senescence syndrome (Biswas and Ghosh 1999). Octopus death can be delayed by delaying its first and last pregnancy.

Method

Some experiments have been carried out on hormones and some interesting results have been collected (Lal 1980, Mukherjee 1978, Mondal 1989). Such as male hormones accelerate senescence whereas female hormones delayed senescence. Some interesting examples are (1) Among differential twins the female one survive longer than male due to their female hormones. (2) Among identical male twins castrated one (goat, ox, man) survive longer compared to their respective same aged normal males. (3) Forcibly castrated males, hermaphrodite by birth also survive longer. (4) Among identical male twins if one voluntarily transexualised into female then she exists longer. (5) But any person in his normal life if loses his reproductivity his death comes soon. Senescence signal may come from the reproductive organs. (6) The women who have beard, their life span is short and the beardless men survive long.

Results and Discussion

The external appearance alongwith the immuno-system fades with ageing i.e. during the onset of senescence. Here the culprit is the DHEA hormone and pituitary hormones (growth hormones) whose levels in the body decline with age. DHEA therapy can prolong the human life to some extent. Natural dietary antioxidants like Vitamin-C, Vitamin-E and β -carotene - precursors of Vitamin-A enhances life.

Many approaches which are very effective for prolonging life are dietary habit will be vegetarian with minimum but sufficient diet which will vary from man to man (2 times under diet and 3 times slight diet) is recommended. Restricted compounds delay senescence. Eating properly in a hygienic way with balanced diet and mild regular exercise can help you get more years out of life.

Avoiding environmental hazards, introvertness, idleness and physical as well as mental and social stress are the passports to a longer life. Agile mind keeps the body fit giving an active life. So frankly mix among the friends and make yourself sociable.

A great deal of information have been available to link dietary habits with ageing such as - malnutrition accelerate senescence which includes food with high in sugar and fat but low in vitamins and minerals. Malnutrition exerts fatigue, insomnia, reduce immunity power, loss of appetite and body weight.

A balanced food which is beneficial for human life span is with high fibres and vegetable proteins, low cholesterol, minimum salt and optimum in minerals and vitamins (fresh vegetables and fruits). Vitamin-C retard senescence. It is proved that sugar accelerate senescence or cause premature ageing. Regular meditation for half an hour or more is the best process to achieve for a longer life found in case of Munis (Roy 1977).

Angry with often aggressiveness negative and evil thought alongwith distressed condition can terminate the life whereas the persons who is immersed in jolliness and good creative thought alongwith social and emotional help from surroundings stay longer in the earth. Even the pet animals enhanced your life.

Gerontologists guess that lower calory intake may either increase the efficiency of cellular metabolism or may act by lowering the blood glucose levels. Restricted food intake reduces the activity of free radicals - the toxic byproducts of normal metabolism that damage cells and tissues of different organs and systems and even the DNA, RNA, protein, catalase.

Human longevity somewhat depend on sufficient pure natural drinking water and on diet low in cholesterol, high in vegetable proteins, rich in natural antioxidants, low in caloric content.

No affluence is essential to sustain active life style and longevity. Younger disease free life is the ultimate goal of our gerontologists.

There are many theories about why we age. It is commonly believed that all multicellular organisms including human repair the wear and tear in their bodies by cell division that renew the tissues. However, as the body ages the cells begin to deteriorate and function less efficiently. Ageing is under genetic control i.e. hereditary because it seems that children of long-lived parents may expect long life span too.

There is an accumulation of toxic or deleterious products within the organism. There is a loss of some essential substance from the organism. Through mutations certain cells accumulate deleterious genes (genetic damage).

In lieu of sweets the bio-sweets and bio-tea can be used to retard senescence. The bio-sweets are completely free from fats and oils. It is free from artificial flavours. Moreover, it has a very high fibre-content, improve digestion and prevent constipation. It is low in calories. The bio-sweets made by acetobacter in coconut milk or in fruit juice is found to be safe for use in humans and animals. These sound sweets are very effective for the figure-conscious fitness buffs as well as to the diet conscious diabetics. Bio-tea is rich with

Vitamin-E like antioxident which prevent cancer. Natural non-caloric intense sweeteners help in reducing the body weight, control obesity and diabetes. Fasting also eliminates cancer producing cells. Fasting saves the energy for cell division and increase the activity of intestine for nutrient absorption.

By genetic manipulation with sugar maple tree (*Acer saccharum*) of United States it was possible to make annual onion into perennial transgenic plants. A scientist of Japan (Makhato-Kureau et al., 1997) discovered a gene called Klotho gene which delay the senescence process in human. As there is the presence of long - lived gene in tortoise in case of animal and in *Ginkgo biloba* and in *Pinus linganeva* (a tree having reached an age of 4,900 years). So, it is not far from to achieve longest life by genetically engineered system in near future.

2.1.8 Herbal Folk Remedies of Bankura and Medinipur Districts, West Bengal

Twenty two medicinal plants belonging to nineteen families used by the tribals / local communities of Bankura and Medinipur districts, West Bengal have been reported as the potential drugs against twenty common ailments of the people.

Keywords: Herbal folk remedies, Tribals, Bankura, Medinipur.

Herbal medicines are assuming greater importance in the primary health care of individuals and communities in many developing countries. There has been an increase of demand in international trade because herbal medicines are very effective, cheaply available, supposed have no side effects and used as alternative to allopathic medicines. Moreover, in most countries the herbal medicine market isn't adequately regulated and the products are therefore unregistered and often not controlled by regulatory bodies. The establishment of regulation and registration procedures has become a major concern in both developed and developed countries.

Two rural districts, Bankura and Medinipur (previously Midnapur, Midnapore) are the tribal inhabited regions of West Bengal. Bankura lies between 22°38' & 23° 38' North latitude and between 86° 36' & 87° 46' East longitudes. It has an area of 6,881.24 sq. km. And the total forest area of 1404 sq. km. Medinipur, the largest district of West Bengal, lies between 21° 36' & 22° 57' North latitude, and 86° 33' & 88° 11' East longitudes. It occupies an area of 14,081 sq. km. and the total forest area of 1,70,895.13 hectare approximately (Basak 1997, Ghosh and Das 1999).

In Bankura, 12% population are tribals. The principal tribals are the Bhumij, Koras, Mahali, Mech, Munda and Santhal. In Medinipur, the tribals are 10% of the total population. Some of the backward groups in this area are Sabar, Santhal, Kadut, Naskar, Munda, Mahato, Sardar, Rajan, Sidhar, etc. Although these people are very poor but they have their own unique culture and life style. Both the tribals and local people of the districts depend on the plants for day-to-day medicinal purposes. Indeed, they have perfected their art of healing too. In spite of rapid urbanization, effecting remarkable changes in the social, cultural and economic spheres, the traditional art of herbal cure and health care is still popular among the tribals and local people. This indigenous system of medicine has almost remained unexplored except for a few reports. The present endeavour aims to record the less known medicinal uses of 22 species of angiosperms as herbal medicines. There are also some earlier reports of herbal medicine applicable to patients in Bankura and Medinipur (Basak 1997, Ghosh and Das 1999, Namhata and Ghosh 1993, Chaudhuri *et al.* 1982, Namhata and Mukherjee 1988).

Methods of Survey

The present work is based on information collected from tribal and local informants concerning common ailments of the people. The survey was conducted during 1999 to 2001, among the villagers, the folk doctors, tribals and local communities. Information was collected from them through personal contact first, and then the treatments were applied to the patients to test their effectiveness. The name of some informers from Bankura and Medinipur districts are: Neul Lohar (Joypur, Bankura), Ratan Kisku (Bachurdoba, Jhargram, Medinipur), Parthasarathi Mishra (Panchal, Bankura) and Amar Kisku (Chilkigarh, Medinipur).

Findings of Survey

The raw materials for medicine preparations were collected from the fields, forests and sometimes purchased from local markets also. Medicines were prepared at home, whenever necessary, by boiling, crushing, mixing the materials, preparing the pastes, etc.

The plant drugs are arranged alphabetically in order of their botanical names; first name is considered in case of multi-component preparations. The plants, their local names, parts used, and method of application against various ailments, have been documented. These are as follows:

Acacia nilotica (Linn.) Willd. ex Delile syn. *A. arabica* (Lam.) Willd. var. *indica* Benth. (Mimosaceae) (Babla).

10 to 12 gram of dried latex, alongwith ghee and sugar, is fried, and used with a glass of milk, once a day, for 7 days in case of spermatorrhoea.

Aegle marmelos (Linn.) Correa ex Roxb. (Rutaceae) (Bel).

The juice extracted from root bark and mixed in a cup of milk is prescribed to induce sleep and as a remedy for depression and weak heart.

Allium cepa Linn. (Liliaceae) (Piyaz).

Bulbs are eaten as raw food especially in summer season to prevent sunstroke.

Amaranthus tricolor Linn. (Amaranthaceae) (Lal-notey).

Root juice is applied locally to the cut wounds to check bleeding.

Azadirachta indica A. Juss. (Meliaceae) (Neem) leaves and *Piper nigrum* Linn. (Piperaceae) (*Golmorich*) fruits.

The leaves of Neem and fruits of *Golmorich*, 5 each, are given to diabetic and high blood pressure patients regularly.

Bryophyllum pinnatum (Lam.) Oken syn. *B. calycinum* Salisb. (Crassulaceae) (*Pathar-kuchi*).

Tender leaves are crushed alongwith a pinch of rock salt and extracted.

Two spoonful extract is taken twice a day in case of hyper-acidity and indigestion.

Calotropis procera (Ait.) Ait. f. (Asclepiadaceae) (*Shet akanda*).

The latex of the plant is locally applied to the cut wounds, till cured.

Citrus sinensis (Linn.) Osbeck (Rutaceae) (*Musambi*).

The patients having general weakness are advised to take musambi juice daily.

Cocos nucifera (Linn.) (Palmae) (Green fruit - *Dab*).

Milk of the green coconut is used as cream to the chicken pox. scar.

Cynodon dactylon (Linn.) Pers. (Poaceae) (*Durba*).

Five shoots are crushed alongwith 25 grains of polished rice in cold water and the paste is eaten once a day in empty stomach for a month for the treatment of habitual abortion.

Ficus benghalensis Linn. (Moraceae) (*Bot*).

Tender *bot* leaves are immersed in half glass of water for overnight and the infusion is used as a drink in the morning every day in case of blood sugar.

Two pieces of prop root cap, fried with ghee, is given to the patient suffering from dysentery, once a day for 3 days.

One drop of fresh latex mixed with a pinch of karpur, is applied externally to the eye in case of cataract, till cured.

Lawsonia inermis Linn. (Lythraceae) (Mehedi).

Pillows made up of flowers are used to induce sleep.

Leaf juice is applied locally to the wounds, 2-3 times a day, to cure candidiasis.

Moringa oleifera Lam. (Moringaceae) (*Sajina*).

One spoonful juice of fresh leaves is taken daily in empty stomach in case of high blood pressure.

Nelumbo nucifera Gaertn. syn. *N. speciosum* Willd. (Nymphaeaceae) *(Padma)*.

Four seeds are crushed in half glass of water and drink is prepared. The drink is given daily in the morning for 21 days for the treatment of habitual abortion.

Piper betle Linn. (Piperaceae) (*Pan*).Warm petiole is applied locally to the cut wounds thrice a day, till cured.

Psidium guajava Linn. (Myrtaceae) (*Peyara*).

Four pieces of mature fruits are immersed in half glass of water for overnight and a drink is prepared. The drink is given every day to the blood sugar patients.

Syzygium jambos (Lam.) Alston (Myrtaceae) (Jam).

One spoonful juice of fresh leaves, alongwith ghee, is taken in the evening in case of enuresis.

Tamarindus indica (Linn.) (Caesalpiniaceae) (*Tentul*).

Tentul fruit is given daily to reduce cholesterol.

Trichosanthes anguina Linn. (Cucurbitaceae) (*Chichinga*).

One spoonful leaf juice is taken twice a day for 10 days in case of menopausal problems.

Trichosanthes dioica Roxb. (Cucurbitaceae) (*Patol*).

Juice extracted from the roasted fruits is used as oil to the chicken pox scar.

Vitex negundo Linn. (Verbenaceae) (Shet nishinda).

Crushed root bark with cold water is given to snakebite patients.

Discussion

It is observed that the traditional medicinal plants in India are fast eroding. It is necessary to inventorize and record all ethnomedicinal information among the diverse ethnic communities before they are completely lost. The tribals of Medinipur and Bankura district, West Bengal use locally available plants both ethnobotanically and ethnomedicinally in various forms. Under herbal folk remedies, the role of these plants has become very important. The present report aims at preservation and dissemination of such knowledge practiced by these tribals and local communities. This information may be helpful in further research because the medicines so prepared by these plants are very effective, cheap and supposedly have no side effects.

2.2 FOR ANIMALS

2.2.1 Ethnoveterinary Medicines from the Tribal Areas of Bankura and Medinipur Districts, West Bengal

Ethno-veterinary medicine preparations from different plants used by the tribals / local communities of Bankura and Medinipur districts, West Bengal, against sixteen common ailments of cattle, have been reported.

Keywords: Traditional veterinary medicine, Livestock, Bankura, Medinipur.

Cattle and their products occupy a unique position in the national economy of the country, which is predominantly dependent on agriculture. While the female progenies supply milk, the male progeny continues to be the principal source of draught power for agriculture and rural transport. Milk is consumed as such, and as butter, ghee (clarified butter oil) and cheese. Among other products obtained - meats, hides, bones and hoofs, and the by-products of the meat industry, such as guts, glands and blood, are important. Hence it is imperative that the cattle population remains healthy and productive. In many areas of the country, traditional veterinary practices are quite prevalent.

Both Bankura and Medinipur (previously Midnapur, Midnapore) are the tribal inhabited districts of West Bengal. The tribals of Bankura are 12% and in Medinipur are 10% of the total population of the respective districts. Here, the tribals are mostly Santhals, Lodha, Mahali, Munda, Bhunij, Oraons and Koras. They are associated with cattle as herdsmen. These men, remaining well below poverty level, have their own systems of the herbal medicine practiced since time immemorial. Human involvement in depletion of the plant resources of the nature and rapid modernization of curative systems, have gradually replaced the ancient but effective systems of tribal medicine. The present report aims at preservation and dissemination of the traditional knowledge of such systems of herbal medicine. There have been earlier reports of herbal medicine applicable to cattle in Bankura district (Mukherjee and Namhata 1988, Ghosh 1999).

Methods of Survey

The present work is based on information collected from tribal informants concerning common ailments of cows, buffalos and goats. The survey was conducted during 2000, among the folk doctors, tribals and local communities. Information was collected from them through personal contact. The names of some informers from Bankura district are :- Gurupada Kisku and Ramkrishna Sahis (Danardi, Bishnupur). Gour Lohar (Rajsole, Joypur), Prafulla Mishra and Samar Tudu (Panchal, Sonamuklhi) and Sunil Hembram (Sara, Raipur). Similarly, Bhim Kalsar (Kushpata,

Ghatal), Nisith Ghosh (C.K. Rd), Pachai Kisku (Jhargram), and Baneswr Mandi (Manikpara) were from Medinipur.

Findings of Survey

People collect the raw materials for medicine preparations from the fields, forests and sometimes, purchase from local markets also. Medicines are prepared at home, whenever necessary, by boiling, crushing, mixing the materials, preparing the pastes, etc.

The plant drugs are arranged alphabetically in the order of their botanical names; first name is considered in case of multi-component preparations. The plants, their local names, parts used and method of application against various ailments have been documented. These are as follows :

Achyranthes aspera Linn. (Amaranthaceae) (*Apang*). - Root is hung in tail for placental retention during parturition.

Adina cordifolia (Roxb.) Hook.f. ex Brandis (= *Haldinia cordifolia* (Roxb.) Ridsdale (Rubiaceae) (*Keli-kadam*) and *Curcuma longa* Linn. (Zingiberaceae) (*Halud*). - A paste of equal amounts of stembark of *keli-kadam* and *halud rhizome* is locally applied to the skin where loss of hair has occurred.

Albizia lebbeck Benth. (Mimosaceae) (*Siris*). - A paste prepared from the stem-bark is applied around the wound of ratbite.

Alocasia macrorrhiza (Linn.) G.Don syn. *A. indica* (Lour.) Spach. (Araceae) (*Mankachu*). - A lukewarm paste prepared from the rotted petiole is applied to the swelling of throat due to cold twice a day for 3 days.

Anthocephalus chinesis (Lam.) A. Richard ex Walpers syn. *A. cadamba* (Roxb.) Miq. (Rubiaceae) (*Kadam*). - The juice extracted from the leaves and mixed with old molasses is fed to animals in colic pain.

Bryophyllum pinnatum (Lam.) Oken syn. *B. calycinum* Salisb. (Crassulaceae) (*Pathar-kuchi*). - Tender leaves (7 in nos.) crushed in 50 gm mustard oil, with 25 gm molasses and 10 gm ajowan, are fed, 2-3 times a day, to animals suffering from poisonous insect bite.

Caesalpinia crista Linn. (Caesalpinae) (*Koranju*). - The oil extracted from the seeds is locally applied to the shoulder wound.

Coccinia grandis (Linn.) Voigt syn. *Cephalandra indica* Naud. (Cucurbitaceae) (*Telakucha*) root, *Nerium indicum* Miller

(Apocynaceae) (*Swet Karabi*) root, *Borassus flabellifer* Linn. (Palmae) (Tal) tender leaf, and *Nigella sativa* Linn. (Kalojira) fruits. A mixture made up of 10+2+2+2 parts of these plant parts are fed to the animals suffering from typhoid fever.

Curcuma longa Linn. (Zingiberaceae) (*Halud*) - The powder of the rhizome is locally applied during leech sucking. It helps in blood coagulation.

Datura metel Linn. (Solanaceae) (*Dhutura*) - A paste made of root with 5 gm pepper is locally applied on head region thrice a day for 2-3 days to cure typhoid fever.

Ipomoea aquatica Forsk. (Convolvulaceae) (Kolni-shak). - The stem is fed to cows of snakebite.

Manilkara achras (Mill.) Fosberg syn. *Achras zapota* Linn. (Sapotaceae) (Safeda) and *Papaver somniferum* Linn. (Papaveraceae) (*Afing gach*). - The crushed fruits of safeda and dried latex of unripe capsule of *afing gach* are mixed with a pinch of lime earth and fed to animals suffering from blood dysentery, twice a day in the beginning; later the dose is reduced.

Moringa oleifera Lam. (Moringaceae) (Sajina). - Powder of the seeds is given to cows immediately after snakebite.

Paederia foetida Linn. (Rubiaceae) (Gandhal) leaf, *Vitex negundo* Linn. (Verbenaceae) (Nishinda) leaf, *Datura metel* Linn. (Dhutura) leaf, and *Cissus quadrangula* Linn. (Vitaceae) (Harjora) stem. - A lukewarm paste prepared from these plant parts is applied to the bone fractures / joints pain.

Piper nigrum Linn. (Piperaceae) (*Gol-morich*). - 10 seeds are crushed alongwith a pinch of rock salt and mixed in 100 gm molasses, and fed to cattle 2-3 times a day with rice gruel to treat flatulence.

Solanum melongena Linn. (Solanaceae) (*Begun*). - A paste of the root and pepper (1 root + 3 pepper) alongwith curd is fed to cow, once a day for 2-3 days in case the cow has ingested cast-off skin of snake.

Spondius pinnata (Linn.f.) Kurz (Anacardiaceae) (Desi-amrah). - The stem-bark is fed to cows in case of snakebite.

Tamarindus indica Linn. (Caesalpiniaceae) (Tentul) leaf & *Amaranthus spinosus* Linn. (Amaranthaceae) (*Kanta-notey*) plant. - The leaves of *tentul* and plants of *kanta-notey* are boiled in water with broken rice and a pinch of rock salt is added and then fed to cattle once a day for higher milk production.

Discussion

It is observed from the foregoing account that the tribals of Bankura and Medinipur districts use locally available plants alone or in various combinations to treat some common diseases of cattle. All the preparations are very effective and cheaply available in both districts and used by them as alternative to allopathic medicines. The treatments show no adverse side effect. Further research on these uses / claims on scientific lines may help in developing effective and cheap drugs for animal health care.

2.2.2 Effects of an Aqueous Extract of *Cardiospermum halicacabum* Linn. Seed on Experimentally Induced Rat Paw Oedema

Abstract

The seed of *Cardiospermum halicacabum* Linn. (Hab-e-qilqil) has been widely used in Unani system of medicine, particularly for the treatment of arthritis and rheumatism. An aqueous extract of the seed of *C. halicacabum* Linn. was studied for anti-inflammatory activity in carragenin and formalin induced paw oedema in albino rats respectively.

The aqueous extract of drug was given in the dose of 500-mg/ Kg body weight orally. The measurements of rat paw were taken before and after injection of carragenin and formalin, with the help of plethysmometer ans screw gauge respectively. The inhibitory action of drug upon carragenin oedema was significant ($p<0.001$). After three hours of administration % inhibition of swelling was calculated as 53.36% compared to control. While % inhibitions of swelling in formaldehyde induced oedema was calculated as 56.44% compared to control. Thus *C. halicacabum* seeds can be said, significantly potent anti-inflammatory drug.

Introduction

The seed of *Cardiospermum halicacabum* Linn. (family-Sapindaceae) commonly known as Hab-e-Qilqil has been widely used in Unani system of medicine and in addition to various medicinal uses, it is also used as anti-inflammatory agent for the treatment of arthritis and rheumatism. Recently it was reported as a bacterial against some pathogenic bacteria, antispasmodic (Shukla *et al.*, 1973) and have some anti-venom effect (Latif *et al.*, 1981). No report on its anti-inflammatory activity has been published to date. The present investigation was undertaken to study the effect of aqueous extract of *C. halicacabum* Linn. seed on anti-inflammatory activity in carragenin and formalin induced paw oedema in albino rats.

Methods

The inflammation in albino rats of either sex (140 gm mean weight) was produced, by injecting 0.05 ml of 1% suspension of carragenin (Winter et.al., 1962), and by injecting 0.1 ml of 4% solution of formalin into the planter tissue of right hind paw in rats (Scyle et.al., 1946). The aqueous extract of test drug was given in the dose of 500-mg/kg body weight orally. Whereas, the control group was given only tap water. The measurements of rat paw were taken before and after injection of carragenin and formalin with the help of plethysmometer and screw gauge respectively. The perecenatge inhibition was calculated after the method of Newbold (1963).

Observation and Discussion

Carragenin induced rat paw oedema (Carragenin group) was measured with the help of plethysmometer found to be 0.72 ± 0.035 ml. A significant increase in the volume of carragenin rat paw was observed after three hours in comparison of volume observed prior to the administration of carragenin. The increase in volume was found to be 0.097 ± 0.099 ml. The mean increase in the volume of rat paw oedema in the control group was found 0.25 ± 0.123 ml.

(Table 2.2.1). The mean volume of rat paw in the group pretreated with aqueous extract of *C. halicacabum* Linn. seed (Hab-e-Qilqil) was found 0.6850 ± 0.048 ml and after the injection of carragenin 0.8016 ± 0.068 ml at zero and after three hours respectively. The inhibitory action of drug upon carragenin oedema was significant ($p<0.001$) after the administration at three hours and % of inhibition of swelling was calculated 53.36 as compared to the control group. The inflammation was also produced by the methods of Scylye et.al. (1946).

Table 2.2.1 : Effect of *C. halicacabum* Linn. Seed on Carragenin induced oedema

	Control Group			Pretreatment Group		
No. of Animals	*Mean Vol. of Paw before Carragenin (ml)*	*Mean Vol. of Paw after Carragenin at 3 hr. (ml)*	*Mean Increase in Volume (ml)*	*Mean Vol. of Paw before Carragenin at (ml)*	*Mean Vol. of Paw after Carragenin at 3 hr. (ml)*	*Increase Paw Vol. (ml)*
6	0.72	0.97	0.25	0.685	0.8016	0.1150
SE±	0.035	0.099	0.12	0.048	0.068	0.06

Table 2.2.2 : Effect of *C. halicacabum* Linn. Seed on Formalin induced oedema

	Control Group			Pretreatment Group		
No. of Animals	*Mean Measurement of Paw before formaldehyde (mm)*	*Mean measurement of Paw after formadehyde at 3 hr. (mm)*	*Mean Increase in measurement (mm) after 3 hr.*	*Mean measurement of Paw before formadehyde (mm)*	*Mean measurement of Paw after formaldehyde at 3 hr. (mm)*	*Mean Increase (mm)*
6	3.93	5.53	1.60	4.20	3.27	0.93
SE±	0.50	0.18	0.20	0.99	0.97	0.46

The anti-inflammatory activity of the seeds of C. *halicacabum* Linn. (Hab-e-Qilqil) in the formaldehyde group was conducted. The paw was measured before and after injections of formaldehyde by screw gauge (made in Germany) having the zero error nil at zero hour and three hours. The mean initial measurement of paw in the control group was found 3.93 ± 0.50 mm. A significant increase in the measurement of formaldehyde induced rat paw oedema was observed at three hours. The measurement was found 5.53 ± 0.18 mm. The mean increase in volume of rat paw oedema in group was found 1.60 ± 0.0208 mm.

(Table 2.2.2). The initial volume of the rat paw in the group pretreated with aqueous extract of C. *halicacabum* seed (Hab-e-Qilqil) was found 4.205 ± 0.956 mm and after the injection of formaldehyde at three hours was found 3.27 ± 0.97 mm and the percentage inhibition of swelling was calculated as 56.44 % compared to the control.

Conclusion

The inhibitory action of studied drug upon carragenin and formalin induced oedema were most significant after oral administration at three hour. This effect of drug may be attributed to the presence of some non-steroidal compounds as the aqueous extract given positive test for alkaloid. The mechanism of anti-inflammatory activity of non-steroidal drug is by their action on anterior pituitary by liberating corticotrophin hormone or by their direct action on adrenocortex (Calesmisk and Buetner, 1949). The extract mechanism of action and nature of anti-inflammatory fraction need further investigation.

2.2.3 Herbal Veterinary Medicine from the Tribal Areas of Midnapur & Bankura Districts, West Bengal

Abstract

Ethnobotanical uses of plants in veterinary medicine by the tribals of Midnapur & Bankura districts have been reported.

Midnapore and Bankura are the districts of West Bengal with a large tribal population. They have their own systems of herbal veterinary medicines practised since the ancient time. Due to rapid

Table 2.2.3

Ailments	*Local Name*	*Botanical Name & Family Name*	*Parts Used*	*How Medicine is Prepared*	*How Used*
Swelling throat due to cold	Mankachu	*Alocasia indica* (Araceae)	Petiole	Rotted petiole crushed made into a paste	Locally applied luke warm paste for twice daily for 3 days
Rat-bite	Sirish	*Albizia lebbeck* (Mimosaceae)	Stem-bark	Crushed and made into a paste	Locally applied around the wound
Leech-sucking	Halud	*Curcuma longa* L. (Zingiberaceae)	Rhizome	Crushed and made into dust	Locally applied to the mouth of leech and helps in blood coagulation
Milk production	(a) Tentul (b) Kanta-Notey	*Tamarindus indica* (Caesalpiniaceae) *Amaranthus spinosus* L. (Amaranthaceae)	Leaf Entire plant	Boiled in water with broken rice and add a pinch of rock-salt	Fed to the cattle once a day
Shoulder-wound	Koranju	*Caesaipinia crista* L. (Caesalpiniaceae)	Seed	Crushed and oil extracted	Oil in lukewarm condition applied externally
Flatulence	Golmorich	*Piper nigrum* L. (Piperaceae)	Seed	Crushed 10 seeds with a pinch of rock salt and mixed in 100 gm molasses	Fed the mixture 2-3 times a day with rice gruel

contd...

Table 2.2.3 – *contd...*

Ailments	*Local Name*	*Botanical Name & Family Name*	*Parts Used*	*How Medicine is Prepared*	*How Used*
Colic pain	Kadam	*Anthocephalus cadamba* (Rubiaceae)	Leaf	Juice is extracted	Fed the juice with old molasses
Snake-bite	(i) Desi- Amrah	*Spondias pinnata* Kurz (Anacardiaceae)	Stem bark	-	Fed to the cows
	(ii) Kolmi-Shak	*Ipomoea aquatica* Forsk. (Convolvulaceae)	Entire plant	Cut the stem according to length of the cow (mouth to tail)	Fed to the animal
Blood dysentery	(a) Safeda (b) Afing gach	*Achras zapota* L. (Sapotaceae) *Papaver somniferum* L. (Papaveraceae)	Fruit Dried latex of unripe capsule	Ingredients of a and b are crushed with a pinch of lime-earth	Fed twice for the beginning days and then doses reduced
Loss of hair	(a) Keli-Kadam	*Adina cordifolia* (Rubiaceae)	Stem-bark	Ingredients of a+ b	Locally applied to the skin until cure
	(b) Halud	*Curcuma longa* L. (Zingiberaceae)	Rhizome	Taken in equal amount are crushed	

advancement in civilization, population and moral deterioration, a gradual depletion of the plant resources of the nature is taking place. The present paper aims at scientific cultivation and preservation of the plants as well as dissemination of the knowledge about such medicine. One should realise the importance of conserving wild vegetation which can be utilized for genetic manipulation also in near future. Earlier publications of Choudhuri et al. (1982); Namhata and Mukherjee (1988); Mukherjee and Namhata (1988); Namhata and Ghosh (1993); Ghosh et al. (1996) have reported the use of herbal medicines in Bankura district. This report is an addition to the above reports.

The present work is based on information collected from tribal localities of Jhargram (Midnapur), Tantipukur, Joypur, Bishnupur, Sonamukhi, Barjora, G. Ghati, Onda, Taldangra, Indpur, Mukutmonipur and Khatra villages of Bankura district. The plants, their local name, parts used and method of application against various ailments have been given in Table 2.2.3.

2.2.4 Indigenous Household Human-Animal Health Care Practices Through Ethnomedicinal Animals

Modern veterinary treatment in rural area is very poor both qualitatively and quantitatively. So the people depend on indigenous animal health care practices through ethnomedicines to maintain their livestock. These bio-resources are very effective in healing ailments in low cost. The present paper unveils the indigenous knowledge about ethnozoology.

Keywords : Indigenous knowledge, Ethnozoology, Ethnomedicine, Livestock.

One could say quite easily that if the treatment entailed 100 steps, then indigenous knowledge contributed at least the 70-80 steps and laboratory science contributed the next 20-30 steps. Tribal communities and others have developed their own knowledge base about the flora, fauna and mineral wealth since the time immemorial. The scientist couldn't enter a dense forest, desert or deep sea etc. on his own interest and choose random plants, animals and minerals to start their research for particular kind of drugs until the indigenous people share their special knowledge about their characteristics and the range of their utility. Their documentation, which should be compiled as a National Bioresource Register, will serve for future tool. Indigenous knowledge is the accumulated information, vision and philosophy of life acquired by the local people in each place and country observing the practical effect of everything when they lived in tune with nature. For the survival of every living being we need the sustained practice of the concept of bioindependence. Ethnic people are knowledgeable and their worldview about the sustainable life is now studied in modern context though Harshberger used the term ethnobotany in 1895. The exploitation of the 1K in different countries by the pharmaceuticals and different multinationals has become a reality and a potential target today. Perhaps the first beings that recognized the medicinal value of plants and inspiration of man for song may be birds and animals. Dogs, monkeys, rabbits, cats, tigers and other animals prefer some plants and animals during some special bodily conditions (i.e. Zoopharmacognosy). There are a number of manuscripts of folk songs and medicines. Folklore items shouldn't be collected for fancy. Maximum old people talking in proverbial language. The claws of tiger if worn in neck are supposed to be good for health; Babies with stomatitis are exposed to smoke of wings of barn-owl for cure. All claims made by tribals should be tested for their validity.

Methodology

The survey was carried out during 1999-2003 to collect the indigenous knowledge (I K) about ethnozoological products as medicines throughout Bengal among tribals, folk doctors and local communities of grandfathers and great grandmothers.

Findings of Survey

The ethnomedicines are enumerated in Table 2.2.4. Through questionnaire survey, sometimes medicines purchased from markets and even medicines were prepared at home, whenever necessary were tabulated on various aspects of ethnozoology, the knowledge on the use of the animals. Each entry has English name, scientific name, local name, part used, preparation of drug and mode of administration in concerned ailments.

Discussion

Ethnomedicines are effective in healing ailments, have low cost and are easily made. However, in the advent of population, pollution, poverty, demands of pharmaceuticals and naturopathy etc., the bioresources and natural resources are degenerating posing the threat of disappearance of indigenous knowledge through globalization local cultures are rapidly vanishing. There should be an upsurgence against the theft of indigenous knowledge through the laws of patents. Ethnomedicines are evolved through over fifty thousand years. So killing animals for fun or gain will destroy nature's food chain.

This aspect of health care system needs further research not only to confirm the medicinal value of such animal based remedies, but also to facilitate more ecologically and socially sound development (Horwitz 1988, Neto 1999).

Table 2.2.4 : Indigenous knowledge on the use of animals and their products in prmimary health care system

Ailments	*Sl. No.*	*English Name*	*Local Name*	*Scientific Name*	*Part used*	*Method of preparation and medicinal use*
	Birds					
Cough	1	Bank myna	Shalik	*Acridotheres ginginianus*	Flesh	Roasted flesh is eaten to treat whooping enough
Paralysis	2	Black ibis	Saras	*Pseudibis papillosa*	Blood	Blood is massaged for curing burning sensation nervous disorder, disorder, paralysis and inherent body heat.
Arthritis and bone fracture	3	Hen	Maraghi	*Gallus domesticus*	Blood (V)	Blood is externally massaged on the aching parts of body. Blood is applied externally for healing arthritis and fracturd bones of cattle, goats, etc.
Cough	4	House crow	Kag	*Corvus splendens*	Flesh	Roasted or boiled flesh is eaten to treat whooping cough.
Asthma	5	House sparrow	Charai	*Passer domesticus*	Dropping (V)	Ash of excreta is used for treatment of asthma in children (three times a day with water for couple of days) and calves.

contd...

Table 2.2.4 – *contd...*

Ailments	*Sl. No.*	*English Name*	*Local Name*	*Scientific Name*	*Part used*	*Method of preparation and medicinal use*
Ear pain	6	Peafowl	Moyur	*Pavo cristatus*	Legs	Legs of peafowl are boiled with oil, which is used to treat the ear pain.
Paralysis	7	Pigeon	Paira	*Columba livia*	Blood	Blood is massaged externally to treat paralysis.
Mating call					Dropping (V)	Fed it for heat induction.
Boils					Dropping	Smear on boils for suppuration.
	Mammals					
Cough	8	Bat	Chamchiki	*Cyanopterus sphinx*	Flesh	Raw flesh having blood is rubbed on the external injuries for healing and also eaten to treat whooping cough and asthma.
Fever	9	Buffalo	Mahis	*Bubalus sp*	Blood (V)	Blood is applied externally on the neck of cow to reduce the body heat.
Constipation	10	Camel	Unt	*Camelus dromedarius*	Dung	Dried dung is burnt and ash is applied externally

Table 2.2.4 – *contd...*

Ailments	*Sl. No.*	*English Name*	*Local Name*	*Scientific Name*	*Part used*	*Method of preparation and medicinal use*
						on stomach for treating constipation.
					Milk	Anti-diabetic.
Arthritis					Blood (V)	Blood is applied externally for treating arthritis.
					Bones (V)	Old bones of dead camel are burnt and ash is mixed with water and fed to the animal for healing rheumatoid arthritis.
Urticaria	11	Cow	Gay	*Bos indicus*	Dung	Dried dung is brunt and ash is applied externally to treat urticar.
					Urine	Urine intake alongwith bark of *Streblus asper* as a remedy for filaria.
Wounds	12	Dog	Kukur	*Canis familiaris*	Flesh	Excreta are applied and flesh is eaten to treat wounds.
Pneumonia and bone fracture	13	Goat	Chagal	*Capra indicus*	Blood (V)	Skin used as cloth after applying turmeric powder on it to treat pneumonia and superficial injury.

contd...

Table 2.2.4 – *contd...*

Ailments	*Sl. No.*	*English Name*	*Local Name*	*Scientific Name*	*Part used*	*Method of preparation and medicinal use*
						Blood is massaged externally on the fractured part and tied with cloth bandage.
Asthma and arthritis	14	Hyena	Hurar	*Hyaena hyaena*	Flesh	Fat is applied externally for healing arthritis; is applied on cloth and the dried cloth is fumigated to cure asthma.
Arthritis, Brighty skin	15	Indian wild ass	Ghudha	*Equus hemionus khur*	Blood (V)	Blood is massaged externally for healing arthritis.
					Milk	Fresh milk is antidiabetic and rubbed in skin for human beauty.
Asthma and arthritis	16	Jackal	Shiyal	*Canis aureus*	Flesh	Roasted flesh is eaten to cure asthma and sciatica.
					Blood (V)	Blood is massaged externally for healing arthritis in goats.
	17	Pale hedge hog	Shukar	*Paraechinus microlapus*	Skin	Ash of skin is used for respiratory problems and cold.

Table 2.2.4 – *contd...*

Ailments	*Sl. No.*	*English Name*	*Local Name*	*Scientific Name*	*Part used*	*Method of preparation and medicinal use*
						Skin is fumigated for curing mouth diseases fo cattle.
					Blood	Blood is smeared on the skin for preventing mosquito.
Rabies	18	Tiger	Bagh	*Panthera tigris*	Tongue	Fed it to animals as a preventive.
Respiration	19	Porcupine	Sajaru	*Hystrix indica*	Squills (V)	Squills are fumigated for respiratory problems in children.
Mouth and wound					Fat	Squills and oil for curing mouth and wound diseases of cattle.
Swelling	20	Rufous tailed hare	Khargos	*Lepus nigricollis ruficodatum*	Blood. Tail	Blood is applied externally for healing swelling : the ash of tail is mixed with oil and the prepared paste is applied for curing burning sensation.
Baldness and Alopecia	21	Rat	Indur	*Ratus ratus*	Whole	Animal is roasted and ash is mixed with mustard oil. This paste is

contd...

Table 2.2.4 – *contd...*

Ailments	*Sl. No.*	*English Name*	*Local Name*	*Scientific Name*	*Part used*	*Method of preparation and medicinal use*
						then mixed with the crushed leaves of *Rivea hypocrateriformis* (Fangyel) and flowers of *Azadirachta indica* (Neem) and applied on the bald head. It is claimed that hair starts growing within 3-4 days of application.
Conjunctivitis	22	Sambar	Horin	*Cervus unicolor*	Antler (V)	Powder of antler is applied in eyes for eye ailments of cattle.
	Reptiles					
	23	Indian flap shell turtle	Kachhop	*Lissemys punctatus*	Carapace (V)	Carapace is burnt and ash is used for healing of internal injuries, pruritis and cough : and for healing superficial blunt injuries of cattle.
Gout	24	Spiny tailed lizard	Girgiti	*Uromastix hardwickii*	Whole body	Whole animal is boiled in oil and the oil is applied externally for joints pain and rheumatism.

contd...

Table 2.2.4 – *contd...*

Ailments	*Sl. No.*	*English Name*	*Local Name*	*Scientific Name*	*Part used*	*Method of preparation and medicinal use*
Wound	25	Iguana	Gosaap	*Varanus bengalensis*	Fat (V)	Smeared the oil on wound of cattle.
	Annelida					
High pressure	26	Leech	Jok	*Hirudinaria granulosa*		Sucking of leech in the wound helps to healing the wound and lowering the blood pressure.
Sciatica pain	27	Earthworm	Kecho	*Pheretima posthuma*	Whole (V)	Eaten to treat the sciatica both in ducks and human.
	Arthropoda					
Colic pain	28	Cockroach	Arsola	*Periplanata americana*	Juice (V)	Fried in ghee should be used in colic pain in cattle and asthma in man.
Cold. Cough. Gout. Asthma Mumps	29	Scorpion	Kankra-bichha	*Buthus meroccanus*	....	Whole animal is boiled in oil and the oil is rubbed in skin, edible also.
Gout, Pain-killer, Sea-sicknes		Torentula	Spider			-do-
Colic pain	30	Lice	Ukun	*Pediculus humanus*	Whole (V)	Fed to the cattle a lice with a guava leaf thrice daily.

contd...

Table 2.2.4 – *contd...*

Ailments	*Sl. No.*	*English Name*	*Local Name*	*Scientific Name*	*Part used*	*Method of preparation and medicinal use*
Senescence	31	Red ant. White ant	Pipilica, Termite		Whole	Regular intake of fried red ant and queen termite delayed senescence.
	Crustacean					
Foot and Mouth disease	32	Crab	Metho Kankra	*Potaman koolovensis*	Whole (V)	Fed to the cattle once daily for 7 days a tablet made from grounded crab and flour of ragi millet.
Ear pain	33	Hermit crab	Sanyasi Kankra	*Uca pugnex*	Whole body	Crab is boiled in water and then taken for asthma : also, it is boiled in oil and used externally for curing ear pains.
	34	Sandy shore crab	Belay Kankra	*Mututa victor*	Whole body	Crab is boiled in water and then the inner watery part taken for asthma : also, it is boiled in oil and used externally for curing ear pains.
						Crab is eaten for the treatment of tuberculosis.

contd...

Table 2.2.4 – *contd...*

Ailments	*Sl. No.*	*English Name*	*Local Name*	*Scientific Name*	*Part used*	*Method of preparation and medicinal use*
Conjuctivitis and Cataract	35	Water snail	Gugli	*Viviparus bengalensis*	Water (V)	Put 2 drops of inner water of shell in each eye thrice daily.
Kidney stone and Cancer	36	Walking fish	Shol	*Chunna striatus*	Brain (V) whole body	Rubbed with water and drank to treat the problems associated with kidney stone. Also used to treat the urinary problems in livestock in the same manner. Fish eaten prevent cancer.
Joint sprain	37	Hammer head shark	Hangar	Zygaena *blochii*	Fat	Fat is applied externally for treatinng joints pain.
Gastric ulcer	38	Sea horse	Samudra Ghora	*Hippocampus cuda*	Whole animal (V)	Dried animal is powdered and mixed with the fodder to cure stomach pain of horses.
Egg (aphrodisiac)	39	Fish bones and Molluscan shell				Fed to the ducks grounded bones and shells alongwith rice bran to get more eggs.

V-Indicates the use of animal/animal products for veterinary purpose also.

Table 2.2.5 : Traditional knowledge about Ethnozoology

Ailments	*English Name*	*Local Name*	*Scientific Name*	*Part used*		*How medicine is prepared and used*
Vomiting and Dysentery	Jackal	Shiyal	*Canis aureus* L.	Leg	-	Smoke of leg
Leucoderma	Boa(Russell)	Bora	*Eryx johni*	Flesh	-	Oil is extracted and then smear
Cough	Peacock	Mayur	*Pavo cristatus* L.	Feather	-	The ash of feather is used
Alcohol/drug addiction	Flying fox	Badur	*Pteropus giganteus*	Droppings	-	Droppings are fed to patient.
Back pain	Iguana (Monitor lizard)	Gosap	*Varanus bengalensis*	Fat	-	Rubbing the oil in back pain.
Epilesy	Hyaena	Hurar	*Hyaena hyaena* L.	Brain	-	Fed the Brain alongwith wheat flour
Piles	Rabbit	Khagos	*Leupus nigricollis* F.	Dropping	-	The smoke of dropping is used
Joint pain	Wild Boar	Banshukar	*Sus scrofa* L.	Oil	-	Rubbing the oil.
Veterinary purpose						
Foot and		Khargos		Legs	-	Smoke of legs are very effective in cattle.
Mouth disease	Monkey	Hanuman	*Presbytis entellus*	Tail	-	Smoke
Cardiac problem	Palm civet or Toddy cat	Bham or Katas	*Paradoxurus hermaphroditus*	Flesh	-	Fed to cattle

3. ENVIRONMENTAL AWARENESS

3.1 TRADITIONAL VEGETABLE DYES FROM CENTRAL WEST BENGAL

Abstract

In the present paper the author has tried to unveil the potential value of vegetable dyes collected from the inhabitants of central West Bengal.

Introduction

Vegetable dyes have been in use since times immemorial. The past fifty years of chemical developments have led to severe threat to human life. Synthetic dyes in food, beverages, textiles, etc. are becoming major hazards and can cause allergies, cancer, digestive disorders, asthma and many other diseases.

Most people, who are suffering from these health problems, don't realize that their ill health can be related to the increasing amount of synthetic dyes and poisons accumulating in their bodies either directly or through the food chain. They are not aware that the synthetic dyes which they are using can be dangerous.

Besides, synthetic additives in junk foods is increasing the violence to human bodies. Monosodium Glutamate (Aji-no-moto) is used to enliven the taste of processed foods and fast foods can cause asthma, depression, acute headache, heart-trouble, etc.

Industrial revolution caused undesirable side effects through all over the world. People are too much needy for natural ways of living. Use of biofertilizer to grow food, nature cure-through acupressure, hydrotherapy, natural dyes, green medicines and cosmetics and mud houses are among the new mania that have caught the imagination of maximum men. There is no denying the event that natural products are in most cases safer than synthetics (Dassanayake 1978, McClure 1927, Hooper 1906).

The substances used to colour foods, drinks, clothes, etc. are called dyes. Prior to 1856 A.D. all dyestuffs were obtained from natural materials, mainly either from animal or vegetable matter, most of them being fast primary colours. The artificial or synthetic or chemicals (called mordants) first was discovered in the year 1856 by Sir William H. Perkin. With the same dye bath one can get different colours depending on the mordant that is used. The utilization of natural dyes has significantly reduced since 1856 after the discovery of synthetic dyes.

Yet we can say that days are changing. Because of the less harmful, low cost and easily available nature of dyes, it has become necessary to intensify the research on natural dyes. However, in recent years attempts are being made to revive this old art of dyeing with vegetable dyes, due to the harmful nature of synthetic dyes. On the contrary, natural dyes are considered excellent for their endurance and soft lustrous colouring. Even after a long period of time they retain their great beauty and charm. Besides the by-products of natural dyes don't create any pollution but in turn may be used for other properties. Another important factor in favour of natural dyes is their property, the natural dyes are very soothing to human body particularly the eyes.

Some insects and tiny molluscans are used to produce natural dyes. An insect of Mexico called cochineal was used to yield bright red colour and purple colour is extracted from a tiny Mollusca (*Murex brandaries*).

Raw materials for dyeing are extracted from natural resources like dried leaves, roots, barks, flowers, fruits, seeds of native plants such as the flowers and pods of Kapila, caramel (a brown colour made from sugar), onion rind, leaves of henna, yellow colour from annatto seeds, saffron (an orange dye from the dried stigmas of *Saffron crocus*, sativus (Iridaceae), dried fruits of harda and fruit rind of pomegranate *Punica granatum* (Puniaceae). Red sandal wood *Pterocarpus santalinus* (Leguminosae) are used for dyeing and these natural colours are still widely used in foods.

The soft raw materials for dye like leaves, flowers or fruits can be boiled in water for an hour. For storage it can be made into powder form. The hard materials, such as barks, roots, seeds, etc.

can be ground into powder and dye-liquid is prepared by soaking it overnight and boiling in water followed by filtration. Wood, silk, cotton, drinks are the most suitable for natural dyeing.

Keeping in mind, the paper is presented on natural dyes-the non-violent and healthy alternative used by local people inhabited in West Rardh region situated between the south side of the river Ganges and west side of the river Bhagirathi in West Bengal. As the soil of this region is red in colour so the name has been derived as Rardh. This region is constituted by the districts of Murshidabad, Purulia, Birbhum, Burdwan, Bankura, Purba and Paschim Medinipur of West Bengal. Its area is 27.500 km^2.

Methods

During the ethnobotanical survey these informations have also been collected by author from many weavers and confectioners as they possess inherited knowledge regarding vegetable dyes. Notes on the vegetable dyes are presented as follows:-

Vegetable Dyes

***Acacia catechu*,**

Khayer, (*Leguminosae*).

The dye is obtained from the heartwood when it is boiled with water used in printing and in colouring pulp and paper.

***Carthamus tinctorius* L.,**

Kusum, (*Compositae*).

Red dye is extracted by crushing the flower petals. The dyes used in sweets, biscuits, rice, etc.

***Tagetes patula* L., and *T. erecta* L.**

Ganda (*Compositae*).

Yellow dye is obtained by crushing the flowers. This dye has an anti-decaying property.

***Nyctanthes arbor-tristis* L.**

Sheuli (*Oleaceae*).

Yellowish red dye is extracted from corolla tube juice and used in chocolate, foods, beverages, etc.

Curcuma longa L.

Halud *(Zingiberaceae).*

Raw rhizomes are boiled for half an hour. After drying, good dyes obtained can be used in any food to cure skin diseases.

Butea monosperma (Lamk.) Taub.

Palash *(Papilionaceae).*

Red dye is obtained by crushing the seeds and mixed with lemon juice, used in foods.

Daucas carota

Gajar, *(Umbelliferae).*

Anthocyanin dye is extracted by crushing the root of black gajar which is used in sweet preparation. Seeds are used in snakebite and also act as an anti-fertility drug.

Lawsonia alba L.

Henna, Mehndi *(Lythraceae).*

Extract the juice of leaves for red dye and mixed with the boiled heartwood of cutch alongwith sugar for aesthetic use.

Camellia sinensis O. Kuntze

Tea, *(Theaceae)*

Red dye prepared from leaves just like henna.

Terminalia chebula Retz.

Haritaki, *(Combretaceae)*

Black dye extracted from the fruit juice and mix a pinch of iron salts for black dye and a pinch of alum for yellow dye.

Rubia tinctorium L.

Madder, *(Rubiaceae)*

Extract the juice of roots for red dye for colouring the cloth.

Conclusion

Documentation of indigenous knowledge about vegetable dyes used by weavers and confectioners is essential for conservation of traditional knowledge. The local people should be convinced that the conservation of natural vegetable dyes is beneficial for their livelihood. It would be helpful to encourage people in income generating activities. These vegetable dyes have no adverse side effect.

3.2 NATURAL VEGETABLE DYES FOR TEXTILE INDUSTRY IN SOUTH BENGAL

Abstract

In the present findings the author has tried to disclose the potential value of easily available natural vegetable dyes collected from the inhabitants of South Bengal.

Introduction

Vegetable dyes have been in use since the ancient times. The past half-century of synthetic chemical dyes caused severe dermatic ailments through textiles such as allergies, skin cancer, eczema, itches, asthma, etc.

The patients don't realize that their ailments can be directly related to the dangerous synthetic chemical dyes. To escape from undesirable side effects world's people are now too much needy for natural vegetable dyes in textiles, timber, food, beverages, etc. Prior to 1856 A.D. all dyes were obtained from nature either from animal or plants or soil. The synthetic dyes (mordants) were first discovered by Sir William H. Perkin in 1856.

Natural dyes are less harmful, low cost and easily available (Dassanayake 1978). So it is necessary to intensify the research on natural vegetable dyes. Besides, natural dyes are considered excellent for their endurance and soft lustrous colouring alongwith very soothing to human body particularly to eyes, breast, under wear, etc.

Method

During the ethnomedicinal (ethnobotany and ethnozoology) survey these informations have been collected by author from many weavers (eg. baluchari designers at Bishnupur-Bankura, Sonamukhi-Bankura, Rajagram-Bankura, Caning-S. 24 Pgs., Midnapore). Notes on the natural vegetable dyes are given below:-

Vegetable Dyes

1. *Acacia catechu*, Khayer (Leguminosae). The dye is obtained from the heartwood when it is boiled with water used in printing.
2. *Albizia lebbeck*, Siris (Mimosaceae). Dye is extracted by crushing the fruits for printing.
3. *Butea monosperma*, Palash (Papilionaceae). Red dye is obtained by crushing the flowers.
4. *Camellia sinensis*, Cha (Transtroemiaceae). Red dye prepared from leaves to colour threads and to clean the teeth.
5. *Ceriops decandra*, Goran (Rhizophoraceae). Bark extract is an excellent source of tannin.
6. *Ceriops tagal*, Matgoran (Rhizophoraceae). Bark yields an excellent dye for textile.
7. *Citrus aurantifolia*, Kagzi lebu (Rutaceae). Juice is mixed with Palash flower extract to procure a permanent dye. Its juice is also used to remove any bad colour and procure a healthy bath water during cough and cold in lukewarm condition.
8. *Daucas carota*, Gajar (Umbelliferae). Anthocyanin dye is extracted by crushing the root of black gajar.
9. *Diospyros peregrina*, Gab (Ebenaceae). Fruit extract smeared on the threads to increase its longevity.
10. *Mimusops elengi*, Bakul (Sapotaceae). Juice extracted from bark used in textiles to increase the lustre.
11. *Punica granatum*, Dalim (Punicaceae). Fruit-rind extract is a very effective dye for cloth.
12. a) *Rhizophora apiculata*, Garjan (Rhizophoraceae).

    b) *R. mucronata*, Garjan.

    c) *R. stylosa*, Samudra-garjan.

    The above 3 *Rhizophora* species bark yield a potential source of dye for dress.
13. *Swietenia mahogoni*, Mahogany (Meliaceae). Dye from fruits an effective to textiles.

14. *Tagetes erecta* and *T. patula*, Ganda (Compositae). Yellow dye xanthophyll derived from the flowers.
15. *Ziziphus jujuba*, Kul (Rhamnaceae).
    Z. oenoplia, Shiakul (Rhamnaceae).

    Ash of their twigs coloured the thread though shiakul is gradually extinct.

Conclusion

Indigenous knowledges are the basic pillars of science though modern men neglected them. Indigenous knowledges which revolved through thousand-year experience still exist in the remote villages. Our scientists must be aware about them and will pay more attention to the effectiveness of vegetable dye.

3.3 EFFECTS OF POLLUTED AIR OF DURGAPUR ON SOME MEDICINAL PLANTS

Air pollutants like dust settling down on mulberry leaves, is high, compared to weeds. In mulberry, decreased in the levels of chlorophyll in the leaves and biomass content, defoliation in young terminal branches, senescence of leaves, reduced leaves and fruits occurred earlier than weeds, indicating that mulberry plants can be used as indicators of the presence of harmful air contaminants.

In Durgapur, air pollution has diverse effects on plants. During the past few years, stomach ailments, respiratory and biochemical troubles have risen sextuple in this area and its peripheral villages. A phenomenal increase has been recorded in the death of small fishes, which are found floating in millions in the Damodar river.

Environmental pollution is caused by the discharge of coal dust and fly ash from the units of thermal power station and the emission from steel plants, cement factory and coke-oven burners in Durgapur. The beds in the centre lay under a thick layer of coal dust. In this area, the air pollution is mostly due to the emission of SO_2 from the smelters of Fe, Cu, Pb, etc. from the burnings of coals in thermal power station and from the petroleum products.

SO_2 around major smelters has provided botanists with an excellent opportunity to study the relative sensitivity of native species. Broad-leaved trees are generally more tolerant than needle-shaped leaved trees. When sufficient numbers of plants are damaged by pollution, the entire population and associated vegetation may be destroyed. There are many instances where entire plant communities have been devastated by prolonged exposure to high concentration of SO_2. 7000 acres of once rich deciduous forest in copper basin area of Tennessee was completely denuded and replaced by grassland sps., following the destruction of the native forest sps. The response of grasses to SO_2 varies within and among species (Howl and Woltz 1981). In recent years, it has become apparent that the moderate SO_2 levels prevailing in urban areas of different parts of the world can inhibit substantially the growth of several grasses (Crittenden and Read 1978). SO_2 is highly phytotoxic and its entry into the leaves of higher plants is primarily through the stomata. This gas interferes directly or indirectly with photosynthesis, energy metabolism and hormonal metabolism leading to reduced growth (Bell and Clough 1973).

As a natural pollutant sink, we all know the plants have the capacity to absorb the dust particles of the environment at a certain level (threshold limit), after which it is dangerous to plants and animals. In the context of serious air pollution in the adjoining areas of Durgapur, the objective of this study was to assess the nature and extent of damage occurring in the growth and development of plant community (especially on mulberry).

For studying pollution with mulberry, the plants selected were *Xanthium strumarium* L., *Eupatorium odoratum* L., *Lantana camera* L., *Croton bonplandianum*, *Ipomoea bona-nox* L., *Tridax procumbens* L., *Mikania scandens* Wild.

For each plant, 3rd and 4th leaves were selected and from each plant equal number of leaf discs (ten) were cut by a cork borer and preserved in rectified spirit separately. The deposition of particulate matters were also estimated. The chlorophyll content from the leaf sample was measured spectrophotometrically according to Arnon (1949). After measuring the chlorophyll, the biomass by leaf discs was measured. The amount of deposited dust particles on a given area at a particular time (3h) was also measured. Different sites were selected for the experiment. The same experiments on the same plants of Barjora (15 km away from Durgapur) was conducted as control. The data were analysed from at least three replicates and the whole experiment was conducted in two different years.

In comparison, with Durgapur, in Barjora the dust concentration on leaves is not significant but the chlorophyll

Table 3.3.1 : Dust conc/cm², biomass (mg/cm²) and chlorophyll levels (in terms of OD values) of different plant sps. on 3rd & 4th leaf.

Plants	*Barjora*						*Durgapur*					
	Dust		*Biomass*		*Chl.*		*Dust*		*Biomass*		*Chl.*	
	3rd	4th	3rd	4th	3rd	4th	3rd	4th	3rd	4th	3rd	4th
T. procumbens	NS	NS	1.031	1.211	0.165	0.175	0.54	0.99	0.937	0.975	0.161	0.171
E. odoratum	NS	NS	1.156	1.452	0.125	0.146	0.99	1.08	1.125	1.312	0.112	0.111
C. bonplandianum	NS	NS	1.625	1.513	0.192	0.218	0.27	0.27	1.511	1.512	0.111	0.071
L. camara	0.02	0.03	1.751	1.625	0.165	0.182	1.08	2.16	1.125	1.061	0.161	0.172
X. strumarium	NS	NS	1.125	1.251	0.135	0.135	0.42	0.27	1.031	1.125	0.132	0.116
M. scandens	NS	NS	0.937	0.875	0.142	0.147	0.27	0.45	0.751	0.875	0.122	0.146
I. bonanox	NS	NS	1.461	1.351	0.151	0.151	0.72	0.45	1.312	1.125	0.121	0.151
M. indica	NS	NS	1.715	1.631	0.155	0.181	1.09	2.18	1.003	1.051	0.081	0.095

NS = Not significant.

content of leaves and the biomass in all the weeds including mulberry is high. The study of pollution effect in Durgapur revealed some interesting features. Injury caused by the pollutants occurred first on leaf blade, than stem. The highest deposition (Table 3.3.1) was found in the leaves of *Morus* sp., *Eupatorium* sp., *Lantana* sp., and in *Tridax* sp. The leaves of other plants has relatively small amount of particulates on their surface. The nature of particulates also differ in Durgapur as revealed visually. The deposited matters may block the stomatal pores thus affecting gas exchange, water relations, other physiological processes, decrease chlorophyll as a result decreases OD (Table 3.3.1), thereby decreases net productivity. Special symptoms were found first in mulberry such as necrosis of leaves, defoliation in young terminal branches hardening of floral buds and reduction in fruit size (data not presented). Similar events occurred in mango (Rao 1972). Since plants are sensitive to dust pollution (Bhatnagar *et al.* 1985), the dust and the 'CO' etc. could be factors involved in pollution to cause reduction in the growth and chlorophyll content. The destruction of chlorophyll is most pronounced in *Morus, Croton* and *Eupatorium* sps., though the quantum of deposition isn't very much significant. The cause of such deleterious consequence may be visualized by the fact that some other factor especially SO_2 gas, besides dust particles may play an important role in this connection. As degradation of chlorophyll in SO_2 fumigated plants has been attributed to SO_2 induced removal of Mg^{++} from chlorophyll molecule converting them into phaeophytin.

3.4 SELECTED PISCICIDAL PLANTS FROM SOUTH BENGAL TO CATCH FISH

Abstract

Present paper unveils the techniques of fishermen to catch fishes. Suggestions for future better fish management through piscicides of plants origin.

Keywords: South Bengal, piscicides, fishermen.

Introduction

Inland aquatic ecosystems occupy a peerless status among the protein sources of the country. Due to continued rise in price of fishes, the protein crisis for human beings in West Bengal have prompted to search for better fish management. Though our West Bengal secures the top position for fish production in India for three consecutive years and in the district Bankura, especially the village - Ramnagar stands first position in spawning.

In the modern pisciculture technique, the preparation of suitable water bodies by eradicating the unwanted fishes and other harmful aquatic flora and fauna is a prime importance. Weed management takes place in our country by manually or chemically by weedicides, insecticides, pesticides or piscicides. These types (highly toxic chemicals) of unwanted biota management is costly, harmful to fish culture for a considerable period.

On the contrary, phytotoxins act as narcotic substance or stupidify the fish sps. and other aquatic fauna, also act as organic manure after decomposition where fishery can exist in close harmony with other uses of water.

Plant derivatives fish toxicants cause respiratory choking, haemolysis of the gills and lower the B.O.D. level of water bodies. Exudates (latex etc.) of some plants cause blindness to the aquatic fauna, swims for a while on the surface of water, comes to rest and caught easily by fishermen.

Phytotoxins are extensively used all over the world for eradication of unwanted fishes during prestocking pond management or during poaching. Use of such phytotoxicants in India is still in nonage. The phytotoxins have the good quality that affected fishes are not harmful for consumption.

Piscicides of plant origin such as mahua oil cake (*Madhuca indica*), root dust of *Derris scandens, D. trifoliata,* seed dust of *Croton tiglium, Milletia spicida, M. pachycarpa,* and bark powder of *Barringtonia acutangula, B. racemosa* are in use since 1834 (Chakraborty *et al.* 1972, Naskar and Guha-Bakshi, 1987).

Generally plant derivatives contain abrin, crotin, clerodin, cerotic, cautohoue, glycosides, kranjin, lingoceric, lindic acid, meliatin, oleic, paraisine, quercetin, rotenone, saponin (Babu 1965, Bhuyan 1967, Tyler, *et al.*, 1981, Deb and Banerjee 1987, Waite 1925, Watt 1889).

Only mahua oil cake (plant toxicant) is used for pond management in certain restricted areas of our country. With the rapid development of pisciculture techniques and due to shortage of supply of pure mahua oil cake it has now become necessary to find out different piscicide sources of plant origin.

Methods

During eco-club movement these informations have been collected by me from different fishermen as they inherited these knowledge from their ancestors.

A few families containing a large number of plants in South Bengal regions possess such piscicidal properties are discussed.

1. *Annona squamosa* L. (Ata), Annonaceae.

Shrub growing up to 10'-15' high. Found generally in village surroundings. Seeds contain anonaine (Alkaloid 0.03%). Seeds are reported to be a piscicide and insecticide, the bark is also astringent. Generally flowering takes place during March and fruiting during April to August.

2. *Achyranthes aspera* L. (Apang), Amaranthaceae.

Annual herb, 2'-3' high, leaf, root and seeds are reported to be toxic to fishery plants emerge from the soil during June-July. Flowering takes place during December.

3. *Cannabis sativa* L. (Bengal Ganja), Cannabinaceae.

Erect annual under-shrub, dioecious, height 7'-8'. Female plant is cultivated as a narcotic and drug plant. Plant is grown during June-July.

4. *Opuntia dilleni* (Cactaceae).

A common xerophyte with phylloclade bearing spines, shrub. The latex is toxic to the fishes.

5. *Tamarindus indica* L. (Tentul), Caesalpiniaceae.

A tall sub-evergreen tree. Occasionally planted and frequently found in forests, roadsides and in villages. Seed dusts are used as toxicant for fishes. This chokes the gills, effects the respiration and kills the fish within a short period. Seed husk of 1800-2000 kg/hac at 1m. depth is enough to kill the fishes. Toxic effect persists for 20-25 days.

6. *Papaver somniferum* L. (Opium Poppy) (Posta), Papaveraceae.

Annual herb with large showy flowers, commonly cultivated. Opium is the dried latex or juice obtained from the unripe capsules of this plant. Crude opium contains many alkaloids. An alkaloid known as morphine is obtained from it. Opium has got narcotic action to fishes. Flowering and fruiting initiates from January-March.

7. *Croton tiglium* L. (Bheranda), Euphorbiaceae.

Plant is small evergreen tree or shrub, rarely found as cultivated. Croton oil is a fatty oil obtained from the dried ripe seeds. Croton oil is a yellowish-brown liquid with a nauseating odour, used mainly as a strong purgative. In India flowers and leaves are also used to poison fishes. Seed dust is a strong safely piscicide. The considerable amount of dust has been recommended 50-60 kg/hac. at 1 m. water depth. The poisonous effect of the seed dust remains from 3-5 days. Blooming from September-October.

8. *Euphorbia antiquorum* L. (Tesiramonsa), Euphorbiaceae.

Erect fleshy glabrous shrub or small tree. It is a common hedge with fleshy 3-angled stem or naturalized in the village shrubberies. Milky latex acts as piscicide. The acrid principle of the latex resin causes blindness.

9. *Euphorbia hirta* L. Euphorbiaceae.

Very small weeds of pathways reported to be poisonous.

10. *Jatropha gossypifolia* L. (Lal-bheranda). Euphorbiaceae.

Are common shrubs on roadside or hedges. Reported to be poisonous.

11. *Sesbania sesban, aegyptiaca* (Joyanti), Leguminosae.

Common in the red soil of the West Rardh. The freshly crushed stem bark is thrown in deep stagnant water of the river to stupefy fishes before catching. This mechanism is commonly used by the local fishermen.

12. *Barringtonia acutangula* Gaertn Lecythidaceae.

Hijal is common in lower Bengal and other parts of India. Much branched moderate tree. Rarely found on the river banks. The bark is used in tanning. The bark is thrown in water to stupefy fish before catching. The bark and seed dust can be used about 150 kg/hac. at 1 meter depth for killing the fishes. The toxic effect of these dust initiate within 2 hrs. and lasts for 3 to 4 days. Flowers and fruits are obtained during May to November.

13. *Acacia concinna* DC. (Ban Rita), Mimosaceae.

Twining shrub, sometimes grow on the village thickets and roadside hedges. Pods contain saponin, which is toxic to fishes. Seeds are coming from June to December.

14. *Melia azedarach* L. (Maha-neem), Meliaceae.

Large deciduous tree which is larger than Neem. Sometimes planted as an avenue tree found here and there. Seed dust acts as piscicide by the process of nervous disorder. A dosage of 2000 kg/hac. in water is sufficient to kill the aquatic fauna within 2 to 5 hrs. The poisonous effect persists for 25-35 days.

15. *Melia azadirachta* L. (Neem or Margosa tree) Meliaceae.

Medium tree, commonly found in village surroundings and in forests, roadsides. Possesses bitter taste in bark and leaves. Fruits contain azadirachtin. It has piscicide effect.

16. *Abrus precatorius* L. (Kunch), Papilionaceae.

Found generally on the forest-bushes or climbing over the trees, on the waste lands. Anthesis takes place during July to December. Pinkish flowers, seeds are red, a black round spot (Black eye) is present on the seed. Seed powder contains abrin. It is highly toxic and soluble in sodium chloride.

17. *Derris indica* (Lamk) (Karanja), Papilionaceae.

Medium tree, rarely found near hedges, waste lands, roadsides and railway tracksides etc. Seed dusts act as piscicides and insecticides. Its oil is very useful. Fruit production takes place during July to March.

18. *Derris scandens* Benth (Noalota), Papilionaceae.

Large climber, very common on forest-bushes (near the scrub-jungles). Its roots containing very poisonous rotenone which inhibits the cellular respiration of fishes. Root powder of 100-150 kg.hac. at 1m. depth is used for killing the fishes and the toxic effect lasts for 12-15 days. *Derris* plant was first used as fish toxicant prior to 1834 in Michigan.

19. *Rhynchosia minima* L. Papilionaceae.

Annual herb. Seed powder is reported toxic to the fishes.

20. *Tephrosia candida* DC, Papilionaceae.

Shrub, generally found in the wet places. The leaves and barks are used as piscicide.

21. *Polygonum hydropiper* L. (Pakur mul) (Knob weed), Polygonaceae.

Annual herb, a common weed of damp places. Commonly found on the margins of ponds, canals, ditches and roadsides of the jungles. Juice of the plants are strongly poisonous to fishes.

22. *Clematis gouriana* Roxb. ex DC (Chhagalboti), Ranunculaceae.

A climbing shrub which climbs up by twisting the petiole, rarely found on the wasteland, hilly areas and on the roadside jungles. Strong acrid, toxic plant used as piscicide. Plants begin to produce fruits during July to January.

23. *Nauclea orientalis* L. Rubiaceae.

Small tree with bushy crown. Rarely planted as ornamental plants. Bark powder is poisonous to the fish.

24. *Madhuca indica* Gmel (Mahua) Sapotaceae.

A large deciduous tree. Occasionally planted and frequently found in Joypur, Sonamukhi forest. Mahua oil cake is a strong poison to fishes and also acts as fish toxicant for removal of unwanted fishes and predators. Within 10-15 days the decomposed oil cakes serve as a good manure to the fish ponds. Leaf abscission initiates during March-April. Flowering takes place in the months of March-April and fruit initiation during May-June.

25. *Pterospermum suberifolium* L. (Muchkunda) Sterculiaceae.

Large tree sometimes found in roadsides and jungles. The wood is red, green fruits take 12 months for ripening. Its flowers and fruits are reported to be poisonous to insects (Bed-bugs etc.) and to the fishes.

26. *Nicotiana tabacum* L. (Tamak), Solanaceae.

A small shrub with sticky glandular hairs. Grown in all climatic conditions throughout the year in a well-drained light sandy loam. Nicotine is present in tobacco plant, as the chief alkaloid mainly as salts of organic acids (Anon, 1986). Nicotine present in leaves is higher (2-5%) in comparison to other plant parts. The alkaloid nicotine bears neurotoxic effect (White et al, 1925). Dry leaf dust is sufficient to kill the fishes, other aquatic biota. Earlier work on different biocidal plant extracts on predatory fishes also supports the present findings (Deb et al, 1987). Toxicogenic effects of nicotine last for 3-4 days after which the zooplanktons began to reappear in the water.

27. *Nicotiana rustica* L. (hookah tobacco), Solanaceae.

It is a smaller plant, height within 1', cultivated during cool weather. Mainly grown in alluvial soils. Grown in rotation with maize, wheat, groundnut, etc. Dry leaf dust acts as piscicides.

28. *Cleome viscosum* Vent (Ghetu), Verbenaceae.

Robust shrub chiefly found in wastelands, living in a flock. It contains clerodin which is sufficient to kill the earthworms and tiny fishes within 30 min. It has also been recommended as an effective insecticide.

29. *Clerodendrum infortunatum* Gaertn (Ghetu), Verbenaceae.

Shrub, height 3-4', found in the wastelands, roadsides. The plants contain clerodin, oleic acid. This plant is reported to be toxic to fishery.

30. *Calotropis gigantea* (Akanda), Asclepiadaceae.

Shrub, latex is injurious to eyes.

31. *Casearia elliptica* (Biri), Flacourtiaceae.

Small trees. Fruit juice causes breathing trouble and unconsciousness in fish.

32. *Cuscuta reflexa* (Swarnalata) Convolvulaceae.

Plant juice is used to enemy's pond to kill the fish.

33. *Strychnos nux-vomica* (Kuchila) Loganiaceae.

Small tree. Powder of seeds is used for killing fish.

34. *Thevetia peruviana* (Kolkey), Apocynaceae.

Shrubs. Powder of seeds is used for killing fish.

35. *Trichosanthes dioica* (Patol), Cucurbitaceae.

Cultivated climbers. Powder of roots is used for killing fish.

Conclusions

The mobilization of indigenous knowledge of fishermen is essential for pond, swamp and river management alongwith the development of rural economy as these techniques are cheap, easily available and have no adverse side effect.

3.5 BEWARE OF THE SELECTED HARMFUL PLANTS TO KEEP THE BODY FIT

A thorough knowledge on poisonous plants is important in our surrounding. Information gathered from the old aged men and from the tribals. Primitive men used them for his kill to poison his arrow. Many poisonous plants are now-a-days used in Ayurveda. The harmful properties are due to toxic substances such as alkaloids, amines, tannins, saponins, glucosides etc. (Chopra *et al.* 1949; Viswanathan and Joshi 1983). Harmful plants are of 3 types:-

(i) Plants poison to man and livestock.

(ii) Plants poison to fishes.

(iii) Insect repellant plants.

The object of this paper is to make childs, students and people aware of such plants so that their harmful effect can be easily avoided. This paper will help the reader to identify the harmful plants. The information mentioned here is based on fieldwork.

1. *Abrus precatorius* (Kunch), Fabaceae.

Climbing shrubs; leaves pinnately compound; flowers pale violet; fruits pods; seed red or white in colour with black eyes. Seeds contain abrin. Chewing the seeds caused fatal, vomiting, severe diarrhoea, gastro-intestinal irritation, trembling of hands.

2. *Alstonia scholaris* (Chhatim), Apocynaceae.

Trees; leaves whorled, simple; flowers greenish white; fruits follicles. Latex contains echitamine. Intake of latex causes toxic effect and if latex falls on eye cause blindness. Flowers causes allergy.

3. *Anacardium occidentale* (Kaju), Anacardiaceae.

Trees; leaves alternate, simple; flowers yellow with pink stripes; fruits drupes with nuts. Its oil is corrosive, causing, blisters on skin and juice is injurious to eyes.

4. *Argemone mexicana* (Shialkanta), Papaveraceae.

Prickly herbs with yellow latex; leaves alternate, simple, pinnatifid; flowers solitary, yellow; fruits capsules; seeds like black mustard. It is a native plant of Mexico but naturalized in India. Dishonest merchants often mixed its seeds with mustard seeds for adulteration. Intake of the seed oil causes epidemic dropsy, diarrhoea, and intense pain all over the body. Cattle avoid the plant.

5. *Calotropis gigantea* (Akanda), Asclepiadaceae.

Shrubs, leaves opposite; flowers in cymes, white or purplish; fruits follicles. Latex contains gigantin and resinols which causes fatal after intake. Latex is poisonous to fishes and for human eyes.

6. *Casearia elliptica* (Biri), Flacourtiaceae.

Small trees; leaves alternate, simple lanceolate; flowers greenish yellow; fruits capsules. Fruit juice is very poisonous. It causes breathing troubles and unconsciousness, Santal, Mech, Oraon and Munda use the fruits for poisoning the fishes.

7. *Cannabis sativa* (Ganja), Cannabinaceae.

Herbs with angular stem; leaves alternate, unisexual. Dried flowers and fruiting tops of female plants yield ganja are strong narcotic, smoking causes hallucination and the central nervous system is out of control. The leaves are used as bhang (intoxicating beverage) or siddhi from both male and female. The resinous exudation from female inflorescences also smoked as charas.

8. *Catharanthus pusillus,* Apocynaceae.

Herbs; leaves opposite, lanceolate; flowers axillary, white, solitary or in pairs; fruits follicles. Plant is toxic to cattle causing madness, temporary blindness with rashes all over the body.

9. *Catharanthus roseus* (Nayantara), Apocynaceae.

Under shrubs; leaves simple; opposite; ovate; flowers axillary, in pairs; fruits follicles. It is a native of Madagascar, but at present both wild and cultivated throughout India. It contains indolic alkaloids Vincristine and Vinblastine which act as poison to heart. Intake of plant juice is poisonous. Madagasian people chew the leaves in order to dull the feeling of hunger.

10. *Cleistanthus collinus,* Euphorbiaceae.

Small deciduous trees; leaves simple, ovate; male flowers in raceme. Intake of roots, barks, leaves and fruits act as violent gastro-intestinal irritant. Tribals take root-bark for suicidal purpose.

11: *Clematis gouriana* (Chhagalbati), Ranunculiaceae.

Climbers; leaves opposite, pinnately compound; flowers greenish white; juice of fresh leaves and stems is poisonous and produce blisters.

12. *Cuscuta reflexa* (Swarnalata), Convolvulaceae.

Parasitic climber with yellowish fleshy stems; flowers pinkish; fruits capsules. Plant juice is harmful. It causes depression with vomiting tendency followed by abortion. Intake of plant causes death of the livestock.

13. *Datura metel* (Dhutura), Solanaceae.

Under shrubs; leaves simple, alternate; flowers solitary, white; fruits capsules, spiny. Seeds are highly toxic. Its toxicity can be neutralized by taking the juice of *Oxalis corniculata* (Amrul). Seeds have also narcotic property.

14. *Datura strumonium* (Kalo-Dhutura), Solanaceae.

Under shrubs; leaves simple; alternate; flowers-solitary; blackish white; fruit capsules, covered with spines. Intake of leaves, flowers, seeds cause dryness of the throat, hallucination, the voice is changed and the vision is affected causing a deep coma.

15. *Euphorbia antiquorum* (Tesira monsa) Euphorbiaceae.

Leafless large succulent shrubs; stems 3-4 winged bearing sharp spines; flowers in pedunculate cymes, red; fruits capsules; Latex is injurious to eyes and strong purgative causing diarrhoea. Tribals put stems in toddy and palm juice to keep it in good condition.

16. *Euphorbia nerifolia* (Monsa), Euphorbiaceae.

Succulent shrubs; stem with stipular thorns; leaves fleshy; flowers solitary; fruits capsules. Latex is acrid, purgative, injurious and cause dermatitis.

17. *Euphorbia tirucalli* (Lonkasij, Latadona), Euphorbiaceae.

Stem succulent with terete branches; leaves scaly; fruits capsules. Latex is poisonous, purgative, acrid, irritant, emetic and harmful to eyes.

18. *Gloriosa superba* (Ulatchandal, Bisalanguli), Liliaceae.

Climbers with tuberous rhizome; leaves oblong; flowers solitary greenish; fruits capsules. Tuber causes fatal, gastro-intestinal irritation and vomiting. Tubers used in poisoning the arrow blades.

19. *Jatropha curcas* (Bag-bheranda), Euphorbiaceae.

Shrubs; leaves simple, palmately lobed; flowers yellowish green; fruits capsules. Seeds are purgative causes vomiting and burning sensation in the stomach.

20. *Jatropha gossypifolia* (Lal-bheranda), Euphorbiaceae.

Shrubs, leaves simple, palmately lobed; flowers dark red; fruits capsules. Seeds cause vomiting.

21. ***Laportea crenulata,*** **Urticaceae.**

Shrubs; leaves with small highly irritant hairs; flowers greenish white. In contact, this plant causes burning sensation for a few days and is aggravated with water. Flowers produce allergenic response, sneezing, insomnia and fever.

22. ***Lathyrus aphaca*** **(Jangli-matar, Ban-matar), Fabaceae.**

Herbs; leaves modified into tendrils. Flowers yellow; fruits sickle shaped. Intake of ripe seeds cause narcotic tendency lathyrism which can be prevented by boiling the seeds in water.

23. ***Mucuna pruriens*** **(Alkushi), Fabaceae.**

Twiners; trifoliate; flowers purple; fruits with persistent bristle-like hairs. In contact of the fruit causes intense itching, blister and dermatitis.

24. ***Nerium indicum*** **(Karabi), Apocynaceae.**

Large shrubs; leaves lanceolate in whorls; fruits follicles. Roots, barks and seeds possess Karabin, which is a powerful cardiac poison like digitalin. Like strychnin it also acts on the spinal cord.

25. ***Papever somniferum*** **(Afing-gach, posto), Papaveraceae.**

Herbs; leaves simple and lobed; flowers too much showy and solitary; flowers yellowish white; fruits capsule. Unripe capsules yield opium (latex) which causes fatal, flushing the face and giddiness. Patients feel drowsy with intense sleep.

26. ***Parthenium hysterophorus*** **(Bish-gach, Parthenium), Asteraceae.**

Exotic herb; leaves irregularly dissected; flowers in white capitulum. Pollen causes allergenic response followed by eruptions on skin. Plants cause eczema. Calves and kids fed on this weed suffer on itching and diarrhoea and even die with severe ulcer in liver, intestine and kidney. In drought condition it causes death of the livestock.

27. ***Pedilanthus tithymaloides*** **(Rangchita, Berachita), Euphorbiaceae.**

Laticiferous shrubs; leaves succulent; flowers red in cyathium. Planted as hedge in gardens. Latex and root are powerful emetic, irritant and injurious to eyes.

28. *Plumbago zeylanica* (Plumbaginaceae).

Herbs; leaves simple, alternate; flowers white in spike. Root contains plumbagin-a napthaquinone derivative which is toxic and an abortifacient. Juice of the plant causes blister on skin.

29. *Plumeria rubra* (Golancha), Apocynaceae.

Trees; leaves lanceolate; flowers white with a yellow centre. Latex contains plumeric acid which is purgative and injurious to eyes.

30. *Ranunculus scleratus* (Bondhoniya, Bon-shim), Ranunculaceae.

Herbs; leaves simple 3-partite with long petiole; flowers pale yellow. Intake of plant is fatal. In contact with the plant causes blisters on the skin.

31. *Ricinus communis* (Rehri), Euphorbiaceae.

Shrubs; leaves simple palmately lobed; flowers in spikes; fruits capsule. Intake of seeds is fatal causing vomiting and circulatory collapse.

32. *Salicornia brachiata* (Salt plant), Chenopodiaceae.

This plant contains saloni salt, which is edible. The beneficial components of the saloni salt are more than the marine or rock salt. This plant is injurious to high blood pressure patient and for the fishes.

33. *Semecarpus anacardium* (Bhelwai), Anacardiaceae.

Trees, leaves simple alternate, large, obovate; flowers greenish yellow; fruits drupes. Latex and juice of the fruit causes eczema and produce blisters on the skin.

34. *Steudnera virosa* (Bish-Kachu), Araceae.

Herbs; leaves with long petiole. Intake of rhizome causes fatal.

35. *Strychnos nux-vomica* (Kuchila), Loganiaceae.

Deciduous tree; leaves simple opposite; flowers greenish white; fruit berries, red in ripe. Intake of seeds cause death. Seed-powder used to kill jackal, dog, cat, rat, etc. and tribals used it for poisoning arrow blades.

36. *Thevetia peruviana* (Kolkeyphul), Apocynaceae.

Shrubs; leaves spirally arranged; flowers yellow and funnel shaped; fruits drupes. Exotic plant with milky latex. Seeds are eaten for suicidal thought causing burning sensation in mouth, tingling of the tongue alongwith vomiting and ultimately leads to death.

37. *Tragia involucrata* (Bichhuti), Euphorbiaceae.

Twining herbs with stinging hairs; leaves simple, serrated; yellowish flower. In contact with the skin causes itching and severe irritation.

38. *Trichisanthes dioica* (Patol), Cucurbitaceae.

Climbers; male flowers often in pair. Root contains trichosanthin. Roots are eaten for committing suicide and is used for killing jackal, hare, deer, cat, etc.

39. *Tylophora indica* (Anantamul), Asclepiadaceae.

Climbers; leaves simple; flowers purple. Juice of the plant contains tylophorine which causes vomiting.

40. *Zanthoxylum aromatum* (Tambul), Rutaceae.

Prickly shrubs; leaves imparipinnate; flowers yellow. Root and stem bark powders are used for poisoning arrows blade.

Conclusion

In West Bengal the following tribals are present which are the chief source of knowledge about ethnomedicine:-Baiga, Bedia, Bhumij, Bhutia, Birhul/Birhor, Chakma, Chero, Garo, Gond, Gorait, Hajang, Ho, Karmali, Kharwar, Khond, Kisan, Kora, Kerwa, Lepcha, Lodha, Lohra, Mag, Mahali, Mal paharia, Metch, Munda, Mru, Nagesia, Oraon, Parhaiya, Rabha, Santal, Sanria paharia, Savar.

A knowledge on harmful plants is essential as some of them are used in ethnomedicine. Information on some of the plants have been actually gathered from the tribals. The present paper will help the common people to identify such plants. The information furnished here are need further research against different ailments of human and veterinary.

3.6 ACUPRESSURE-A WAY OF NATURAL LIVING

Acupressure is The Simple Treatment to Perfect Health *No Drugs, No Puncturing, No Side Effect

1. We are left spellbound by the wonders of the world like Eiffel Tower, Empire State Building, The Taj Mahal created by man, forgetting that the greatest wonder on this Cosmos is the man, its human body.

   Our body is equipped with the best, automatic, delicate but most powerful machines, viz. Heart and lungs-nonstop pumping sets, Eyes-wonderful camera and projector, Ears-astounding sound receiving system, Stomach-wonderful chemical-laboratory, Nerves-miles of communication system, Mind-unparalleled computer with infinite capacity; and greatest thing about it is the unbelievable coordination of all these machines so that this body can easily work for even more than 100 years.

   Now, it is observed that in any good machine, provision is made whereby it automatically stops when there is danger, e.g. Refrigerator and Geyser etc., which restart when you push its switch. Then there is no surprise if such provision is made in the human body and to energize the malfunctioning organs. This health science is popularly known as Acupressure. This therapy is the most precious gift to Mankind by the Creator Himself.

2. This therapy was known to our ancestors in India even before 3000 years. Research has been made in USA and this is found to be very useful. Many doctors use it in their practice. I learnt this therapy during my visit to Siliguri in 1994.

   This therapy is tried on many friends and others including patients of Heart Attack, Paralysis, Asthma, etc., and results are astounding. This therapy is found effective on all types of diseases, including even Cancer.

   This is the only therapy where you become your own doctor, have daily medical check-up, can treat yourself and even prevent diseases. There are no side effects. This is a DO IT YOURSELF therapy.

3. According to this therapy, electric current-Chetna is passing in the body as per wiring lines shown in Chart I.

   So if the current isn't reaching any part of the body due to any fault, there is pain or disease in that part. So if proper current is sent there on that spot, the pain should subside and disease should be cured. The switchboard of this current is in two palms and two soles of the legs.

4. According to this therapy of Acupressure, treatment means giving pressure or massage to these points and around them on two palms or soles only. This pressure can be given by pressing thumb and first finger or by unsharpened pencil, etc. on the points for 4 to 5 seconds and let go for 1 to 2 seconds and then repeat it for 1 to 2 minutes like pumping. This way treatment is to be taken 3 times a day for any disease till it is cured.

5. Body is divided into two parts. For anything wrong on right hand side, treatment is to be given on right hand, palm and vice versa. Further, it is subdivided into front and back for all parts situated in front of body like eyes, lungs, etc. pressure is to be given on the palm / sole and for organs like spine, etc. situated on the back part of body, pressure is to given on back of the palm / sole. On points of Endocrine glands, deeper pressure should be given by pressing the thumb vertically.

   For any other problems connected to organs not shown in ready reckoner, pressure is to be given on tips of fingers and toes relating to that part of the body covered in chart 1.

6. For complete health and to keep the body fit and prevent diseases, this treatment should be taken daily 5 minutes for all the points on one palm and another 5 minutes for all points on other palm.

7. While doing this Acupressure treatments if you feel pain at any points, it means fuse has gone from that part and disease has started in that organ relating to that point. Thus the disease/pain is reflected on the point of palm and to cure the same treatment must be taken 3 times a day for 2 minutes like pumping till pain on the point is gone. The motto is, if you feel pain, press it out. By having treatment daily, all the organs will be re-charged and with this daily

medical check-up, anything wrong in the body will be found out immediately and cured with immediate treatment. That will ensure against possibility of any disease like heart attack, paralysis, blood pressure, diabetes, cancer, etc.

This treatment can be taken at any time during 24 hours. The treatment can be taken by oneself or can be given by anybody else, e.g. children or invalids can be given treatment by parents or others. It should not be taken within 1 hour after meal.

8. **For nervous tension:** Clasp your hands tightly interlocking the fingers. Then with left hand fingers, press once back of the right hand and then with right hand fingers press on the back of left hand, to be repeated 1 to 2 minutes, 3 to 4 times a day. This treatment can be to ensure good sleep and cure insomnia.

9. **Endocrine Glands:** These glands point No. 3-4-8-14-15-25-28 & 38 play a great part in governing the proper functioning of all the vital organs and maintaining the metabolism of the body. We know that due to less functioning of Pancreas, we would suffer from Diabetes. However, we don't know that due to overworking of this Pancreas, we get low B.P. dullness, migraine headache and leading to alcoholism. Same way Adrenal gland promotes keenness of perception, untiring activity, inner energy, courage and vigour. Further, they intensify the flow of blood and control Blood Pressure. This therapy only teaches us to control all these glands. Fifteen days regular treatment on these points increases the will-power to stop bad habits and reduces the side effects when bad habits are stopped.

10. **Anaesthetic effect:** If continued pressure on any point is given for more than 3 minutes, it creates anaesthetic effect. This is useful in case of severe pain in head, toothache or any other place. Care should be taken not to overdo it and do many times.

11. **Diabetes:** Over and above pressing points 25 and 26, drink the following water. Take plenty of coriander leaves. Extract half a cup juice and drink it first in the morning for 20/30

days. Do not take breakfast for 30 minutes. Avoid anything for 30 minutes.

12. **Diseases of stomach:** Disorder of digestive systems-gas etc. In such cases, take a wooden rolling pin with horizontal lines in cut (velan used for making papad), put the same on ground, put the legs on it, roll it for ten minutes, avoid heavy food, drink more of health drink.
13. **Skin diseases:** It is not an independent disease. This disease means the body is trying to throw out the poison created by wrong foods and toxic drugs and can't be cleared through urine. So drink more of health drink, avoid such foods and drugs and give treatment on points 11, 18, 23 and 26.
14. **Chronic diseases:** In such cases and for treatment of bed-ridden people the following Health-drink is very useful. Mix 300 gms of Amla powder with 75 / 100 gms of ginger powder. Take 1 teaspoon with water twice a day. This drink is useful for all the people.
15. This treatment should also be taken by even healthy persons to ensure good health and physical fitness and to prevent any disease. The switchboard of the palms has not been used until now; so there may not be response for 2-3 days. However, continue treatment for 10-15 days to know the results. This is a Science and not a subject of belief, so results will definitely be obtained. During the last 14 years over 9 million people have got benefit from this therapy in India.

A healthy answer, the natural way!

Bringing sustainable and chemical free medicine for protecting your health and nature's diversity having no side effect.

Disease	**Points**
1. Cold, Tonsils	1 to 7 & 34
2. Cold, Cough	1 to 7, 30 & 34
3. Sinus	1 to 7, 34 and Tips of all fingers and toes
4. Fever due to cold or Bronchitis	1 to 7 & 34
5. Fever due to Malaria	1 to 7 & 37
6. Headache due to Cold	1 to 7 & 34

7. Due to heat and Migrane	1 to 7 & 34 also 22, 23, 25 & 27

Disease of Stomach

(In the case of such problems silver water is useful. Correct Solar Plexus.)

8. Pain in Stomach lose motion vomiting	19, 20, 22, 23, 27
9. Gas Trouble	19, 20, 22, 23, 27
10. Jaundice	19, 20, 22, 23, 25, 27, 28, 38
11. Acidity/Hyper tension	22, 23, 25
12. Appendicitis	21
13. Constipation	Press middle of chin 3 to 6 minutes
14. Allergy	21 calcaria phos 12 × 4 tables twice a day
15. Stone and urinary disease	11-15, 18, 26+silver water
16. Fatigue	32+Drink like warm water
17. Epilepsy-fit	All points
18. Anaemia	37
19. Piles	10+Treatment for Constipation Drink more water
20. Hernia	11
21. Falling of hair	Rub nails of 8 fingers against each other for 10/12 minutes.
22. Pain legs	Roll your feet of five minutes grooved wooden roller. Give treatment on points of sciatica nerve.

All points are in palm and upto one inch below the wrist. So in order to maintain good health and prevent disease, give treatment on both the palms or soles for five minutes daily. In case of any disease, treatment is to be given on the points of two palms and/or soles shown against each disease. Do not worry, if no results within 2/3 days. Continue this treatment.

P.S : Pressure is to be given on and around all the points shown in the palms or soles. (refer book)

23. Blood pressure (Low)	3, 4, 22, 23, 25, 28
Blood pressure (High)	3, 4, 8, 14, 15, 25 & 28

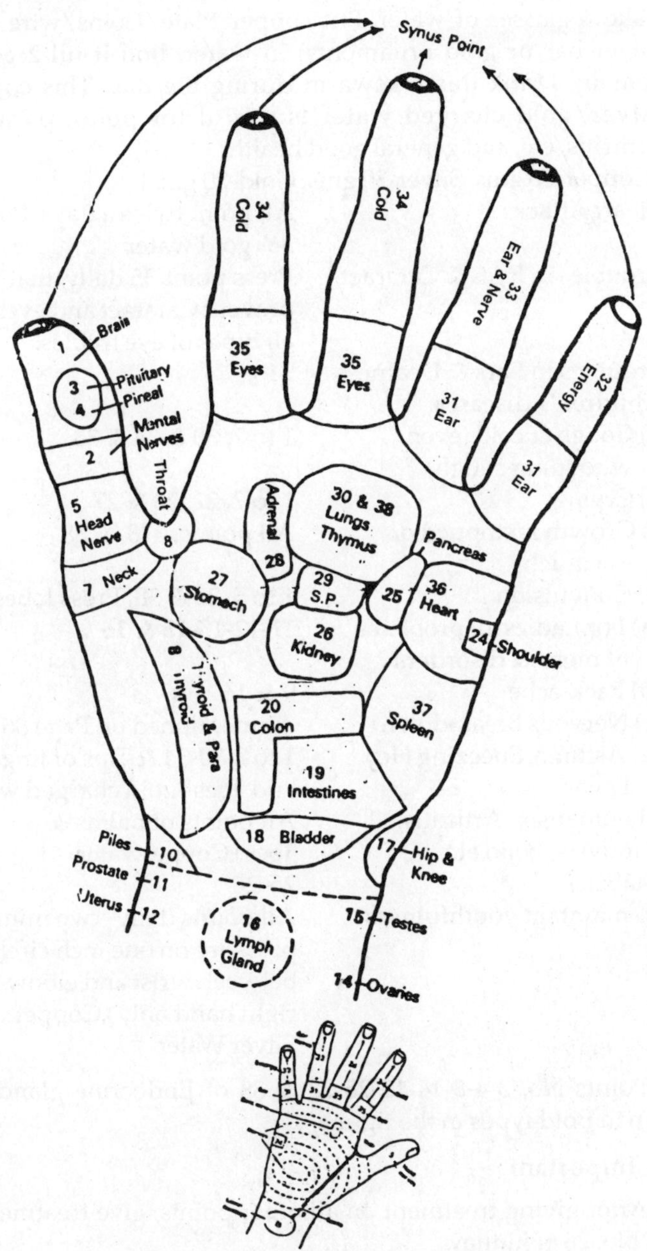

Left Hand showing different points

Take 4 glasses of water. Put copper Plate/Coins/wire (even silver bar or gold ornaments) in water, Boil it till 2 glasses remain. Drink this lukewarm during the day. This copper/silver/gold charged water is useful for polio, paralysis, arthritis, etc. and general good health.

(Copper-60 gms, Silver-30 gms, Gold-20 gms)

24. Heart Attack	All point twice a day+ Point 36+gold water
25. Troubles in Eyes & Cataract	Press point 35 daily, that will prevent Cataract and even cure all types of eye trobles.
26. Troubles in Ears & Deafness	31
27. Children's diseases:	
a) Cough & cold (even whooping cough)	1 to 7, 30 & 34 & 38
b) Crying	1 to 7, 22, 23 & 27
c) Growth is stopped or too much	All points, 3, 8, & 38
d) Convulsion	1 to 5, 28 & 38, Press lobes
28. (a) For Ladies all problems of mens & disorders	11-12-13-15 & 16
(b) Back-ache	9 & 16
(c) Nervous Break-down	As mentioned on Page 86
29. (a) Asthma, Sneezing Hey Fever	1 to 7-30-34 & Tips of fingers and Toes+gold charged water
30. Rheumatism, Arthritis, Paralysis, Polio etc.	All points of palms & Toes+Copperwater
31. Diabetes	25-26
32. To maintain youthfulness	All points daily+two minutes pressure on one inch circle between wrist and elbow on right hand only+Copper Silver Water

Points No. 3-4-8-14-15-25-28 & 38 of Endocrine glands are shown in Bold types in the figure.

Very Important

After giving treatment on all other points, give treatment on point No. 26 of kidney.

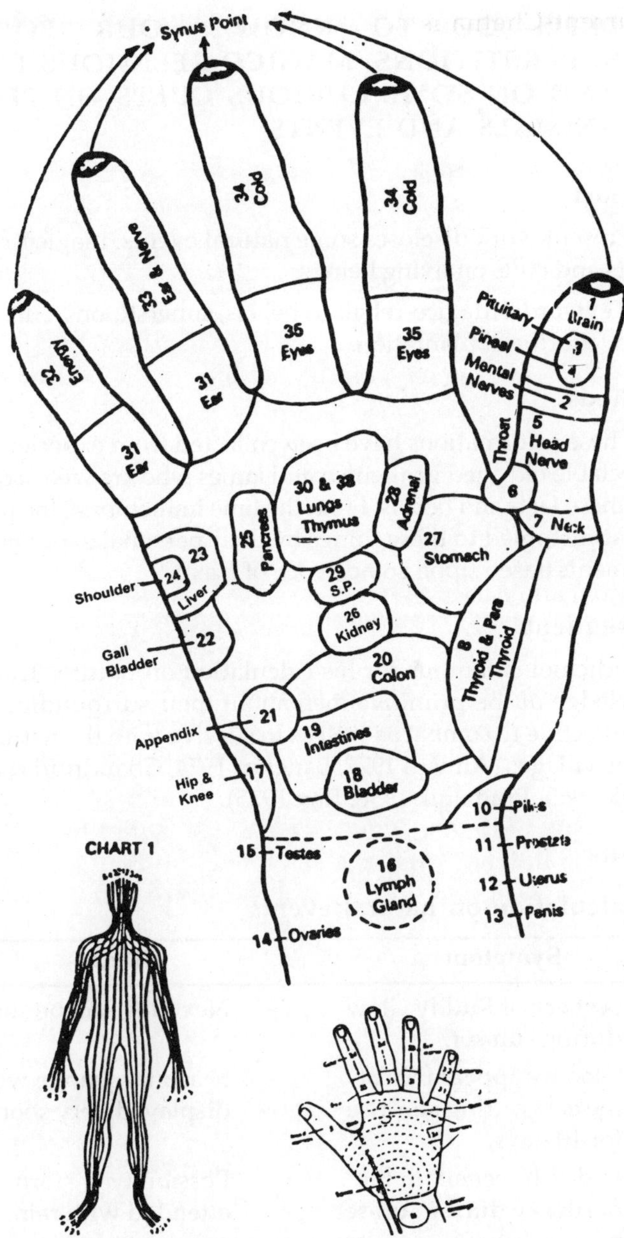

Right Hand showing different points

3.7 LET'S GO TO KNOW OUR POPULAR SUPERSTITIONS, MAGICO-RELIGIOUS BELIEFS AND ON SOME CURIOUS CULTS ON PLANTS, ANIMALS AND EVENTS

Abstract

Present work discloses some natural events, magico-religious beliefs and cults on living beings.

Keywords: Magico-religious beliefs, superstitions, cults, house-lizard, courtyard plantation.

Methods

These informations have been collected from experienced and respectable old aged gentlemen and ladies who are well acquainted with these facts and beliefs. From the time immemorial the primitive men accustomed to these things by their personal experience. The arguments based upon coincidence of events.

Consequent

Jadichiel is famous for his calculation on nature. Traditional knowledge of the primitive men about their surroundings is still very effective to combat as well as to survive with the nature in the remote villages (Amiadi 1977, Banerjee 1974, Chaudhuri *et al.* 1975, Nayak 1965, Pramanik 1956, Roy 1925).

Results

A. Calculation on natural events

	Symptom		Effects
1.	Scenery of Ruddy Glow during sun-set.	-	Next day will be sunny.
2.	Gloomy appearance of the east during sun-rise for 10 days.	-	Seasonal change will be displayed very soon.
3.	Reddish scenery in the north sky during sun-set.	-	Possibility of storm attended with rain.
4.	Beauty scene surrounding the moon at night.	-	Rainy day will come soon.

5.	Downward movement of smoke from woven-chimney.	-	Rain will come soon.
6.	Heavy fog-fall.	-	Day will be sunny.
7.	Fickle-minded condition of dog-cat alongwith cry out.	-	Earthquake or natural startling event will occur.
8.	Cry out of swallow or crane.	-	Rain will come soon.
9.	Herd of cows running to and fro.	-	Thunder-storm will happen.
10.	Suffocation feeling during summer.	-	Immediately rain will appear.
11.	Aquatic birds flying to and fro over the swamps and tanks having no efforts for food intake.	-	Possibility of storm attended with rain.
12.	Appearance of migratory birds (Herons) from Siberia.	-	Pre-indication for rainy season.

B. On the eve of departure for journey, examination, competition, noble work, etc. to achieve success following are the good indicators if you seen :-

(a) Jackal from left side crossed your path.

(b) Snake from right side crossed your path.

(c) Human dead body carried for cremation ground.

(d) Large *Labeo rohita / Catla catla*.

(e) Wear marks of curd-paste in your fore-head.

(f) Filled vessel.

(g) Journey at dawn of Tuesday.

(h) Fortunate owl (barn-owl) / a pair of Indian Myna (*Acridotheres tristis*).

(i) Flying white (fairy) kites in the sky / low moans of a pair of dove.

(j) Touched towel during the time of intercourse of cobras.

C. Following are the bad indicators i.e. opposite to the above:

(a) Empty vessel.

(b) Egg.

(c) Banana.

(d) Behind call of a person or croaks of crows.

(e) Very ugly owl (strix / screeches).

(f) Inauspicious portion of the day specially Thursday.

(g) Inauspicious portion of the night.

(h) Vibration of the right eye of a woman.

(i) Evil omen of sneezing and the chirping of a house lizard.

(j) Cat crossed the way of a vehicle driver.

(k) Flying vultures in the sky, stumbling or sudden hit in body.

(l) Cry out or croaks of crows.

(m) Cry out of dogs and cats.

(n) Iguana in the way.

(o) Eunuch seen in unwashed mouth.

(p) Beardless man or woodcutter seen in the way.

D. Effect of down fall of house lizard in different parts of human body

Sl. No.	Parts		Effects
1.	Head	-	Will be rich
2.	In ear	-	Gain ornaments
3.	In rib and eye	-	Meet a friend
4.	In nose	-	Gain aromatic things
5.	In mouth	-	Gain sweets
6.	In neck or palm	-	Have some money
7.	In arm	-	Achieve happiness
8.	In back	-	Achieve land
9.	In anus	-	Endanger life
10.	In thigh	-	Loss of money
11.	Movement of lizard from foot to upward direction	-	Bad indication
12.	In front of	-	Gain vehicle

E. Bengali superstition regarding courtyard plantation

Name of plants and location		Effect
1. East side of your house with *Michelia champaca*	-	Have no scarcity of money
2. Mango, jackfruit, lemon, wood apple, kul	-	Have sufficient tenants and sons
3. South-east side of the house with pineapple and black-berry	-	Flourished with kith and kin, friends
4. Betel-nut tree in south-west side of the house	-	Increased the offsprings
5. Coconut tree in the surroundings of the house	-	Though possess good sign for family but caused thunder
6. Palm tree (*Borassus flabellifer*) tamarind, fig, shimool, banyan tree, bamboos	-	Have no prosperity of the family.
7. *Streblus asper* (Shewra), (Moraceae)	-	a) Protect the house from thunder b) Premarriage with the tree protect the man from widowness c) It leaves as a fodder decreased the yield of milk in goat

F. Some popular misconceptions and cherished beliefs

Wearing of a gold ring cures a stye, a painful inflammation of the eyelid.

People claimed that wheat germ can prevent ageing as well as make a person more sexually potent.

The meat-eaters have extra strength and stamina is a myth. The nutritional inputs of a vegetarian can be favourably high compared with those of a meat-eater.

Swallowing a raw egg is a daily ritual mainly because people believe that raw egg is more nutritious than cooked. This is not true. In raw condition the white egg albumin contains an antidigestive factor which can be spoiled only on heating.

Fishes are the brain tonic but there is no reason to believe that a fish-eating population is not in any way superior to that of non-fish eaters.

The people with digestive problems should consume sour curd everyday to cure digestive problems is a popular myth. For those persons who have an allergic reaction to lactose in milk, curd may be helpful because it is easily digestible.

There is a general belief among the drivers that a black cat crossing the path brings bad luck or carrying a rabbit's foot will bring good luck.

Passing salt, knife, fork or utensil on the floor means guests are coming.

Daily intake of multiple vitamins and mineral supplements is also a fallacy because all necessary nutrients are easily obtained from a sensible balanced diet.

Everyone likes bright and pretty skin colour but evolution favours darker skin colour because melanin blocks harmful ultraviolet rays in sunlight. That is why light and fair skin coloured people are more susceptible to skin cancer than darker people. It is proved that people with darker eyes have faster reaction times than those with light coloured eyes. The extra pigment in the darker eyes speeds up transmission of nerve impulses from the eyes to the brain. As a consequence people with darker eyes would be a good sportsman compared to the fair skin colour.

G. Magic oriented superstition

1. Irony nail used to exorcise ghost.
2. To escape from evil eye collyrium is smeared on eyelids of children or kept custard-apple leaf in the ear. Care-free napkins of females kept far from evil men.

H. Tabu-based superstition

1. Unwashed mouth after rice intake creates acne in the face.
2. Pregnant women must be avoided the followings to get normal child :-

   Don't eat twin eggs or twin bananas or tear off betel-leaf or eat during the time of eclipse.

Face nature, Face ignorance, Face illusion.

Never Fly

– Swami Vivekananda

3.8 BENEFICIAL COMMON BIRDS IN BENGAL

Next to mammals birds are the important creatures of the earth. Still birds meat and bones are very popular remedies against various ailments among the tribals.

1. Little Cormorant (Choto Pankouri) ***Phalacrocorax niger***

Size : Domestic Duck

Description: Clistening black duck-like water bird with longish stiff tail, slender compessed bill sharply hooked at the tip. Small white patch on throat, Sexes, alike. Found on all inland waters, also brackish lagoons and tidal creeks. Lives exclusively on fish which it chases and captures undet water. A familiar sight is a cormorant on a rock or a branch crying is outstretched wings.

2. Pond Heron or Paddy Bird (Bak) ***Ardeola grayii***

Size : Smaller than a village hen.

Description : Egret like marsh bird, chiefly earthy brown when at rest but white in flight. Breeding season add a maroon hair-like plumes on back and long white occipital crests. Sexes alike. Found singly or in loose parties throughout India union. Found close to water. Its call is a harsh when it takes off. It eats frogs, fish, crabs and insects.

3. Parian Kite (Chil)

Size : 12-24 inches.

Description : Our commonest reptor, this large brown hawk, distinguished from all similar birds by forked tail, particularly in flight. Sexes alike, Singly of gregariously, scavenging in towns and villages. It eat offal and garbage, earhworms, winged termites, lizards, mice, disabled or young birds. The call is a shrill ewe-wir-wir-wir.

4. White breasted Kingfisher (Machhranga)
State Bird ***Halcyon smyrnesis***

Size : Between myna and pigeon.

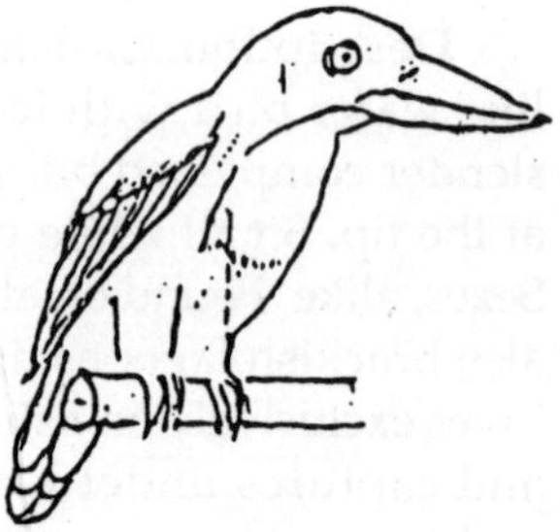

Description : A brilliant turquoise blue, with deep chocolate brown head, neck and underparts. Conspicuous while "Shirt front" and long heavy pointed bill. White wing patch conspicuous in flight. State bird of West Bengal. It eats fish, tadpole, lizard, grasshopper and other insects. Occasionally young birds and mice. It has a loud crackling, rolling call. Least dependent on water among kingfishers. Seen at pond, rainfields, beacites and paddyfields.

5. Coppersmith or Crimson breasted Barbet ***Megalaima haemacephala***

(Choto Basanta Bauri)

Size : Sparrow, more dumpy.

Description : Heavy-billed grass green barbet with crimson breast and forehead, yellow throat and green streaked yellowish underparts. Short truncated tail distinctly triangular in flight silhouette. Sexes alike. Singly or loosely in parties. Found on fruiting trees especially wild figs. Familiar loud, monotonous rather metallic tuk tuk call. It feeds on fruits, berries. Especially banyan and peepul figs. Sometimes eats winged termites.

6. Lesser Golden backed Woodpecker ***Dinopium benghalensis***

(Choto Sonali Kaththokra)

Size: Slightly larger than the myna.

Description : Upper plumage golden yellow and black, lower buffy white streaked with black, more boldly on breast. Crown and occipital crimson. It is found on tree trunks in open wooded country, orchards, etc. Partial to mango topes, groves of ancient trees and coconut plantations. Works up stems and boughs directly or in spirals tapping on the bark and chiseling away rotten wood for beetles and insects. Black ants, pulp of ripe fruit and flower nectar are also part of its diet. The call is a loud, harsh chattering "laugh".

7. Indian Myna ***Acridotheres tristis***

(Shalik)

Size : 9" inches.

Description : Dark brown bird with bright yellow bill, legs and bare skin around the eyes. Large white patch on the wings, conspicuous in flight. Lives in close association with humans and eats almost anything.

8. Jungle Myna ***Acridotheres fuscus***

(Jhunt Shalik)

Size : Indian Myna.

Description : Same as Indian Myna, but more greyish overall. Absence of bare yellow patch around eyes. Tuft of feathers on the forehead.

9. Pied Myna ***Sturnus contra***

(Go Shalik)

Size : Indian Myna.

Description : Black and white myna, orange skin around eyes. Deep yellow and orange bill. Sexes alike. Lives in close association with humans though never entering or utilizing houses like the Indian myna.

10. Red whiskered Bulbul ***Pycnonotus jocosus***

(Sipahi Bulbul)

Size : Red Vented Bulbul.

Description : Brown above, white below with a broken necklace on breast. Black upward pointing crest, crimson whiskers. Prefers better wooded locality than the Red vented Bulbul. It eats insect, fruits and berries. The call consists of a variety of cherry notes, more musical than that of the Red vented Bulbul.

11. House crow *Corvus splendens*

(Kaak)

Size : 17 inches.

Description : Black with grey neck which distinguishes it from the less common all black jungle crow. Lives in association with human and eats almost everything.

12. Red Vented Bulbul *Pycnonotus cafer*

(Bulbuli)

Size : 8 inches.

Description : Perky smoke-brown bird with partially crested black head, scale-like markings on breast and back, conspicuous crimson patch below root of tail. White rump noticeable in flight. Sexes alike. Common in gardens and small scrub jungles both near and away from humans.

13. Tailor Bird *Orthotomus sutorius*

(Tuntuni)

Size : Smaller than a sparrow.

Description : Small olive green bird with whitish underparts, a rust coloured crown and two elongated pin-pointed feathers in the tail which is carried jauntily cocked. Sexes alike. Found singly or in pairs, in shrubbery. The call can be described as pretty-pretty or towit-towit-towit. It feeds on insects and their eggs and grub, flower nectar.

14. Magpie Robin *Copsychus saularis*

(Doyel)

Size : Bulbul.

Description : Trim black and white bird with cocked tail, which it waves up and down. Black replaced by grey in female. Found singly or in pairs in and around human habitation. It eats insects picked off the ground and flower nectar. The call is a plaintive swee-ee or chur-r. It is a very good mimic of other bird's call.

BIRD WATCHING TIPS

Bird watching is a very interesting hobby and an important field of study. Birds are beautiful, some have very melodious calls and they have a very important role to play in the life of man. Birds are important checks on pests like insects and rodents; they help in pollination and dispersal of seeds. Scavenger birds help by removing garbage and the droppings of sea birds provide excellent fertilizer! They maintain the ecological balance.

For bird watching you just need sensitive ears, keen eyesight and an alert disposition. A pair of binoculars and a bird guide are helpful aids.

Guide Books

- Book of Indian Birds – Salim Ali-BNHS
- Pictorial Guide to Birds of the Indian Sub-continent – S. Ali D. Ripley-BNHS
- Hand Guide to Birds of the Indian Sub-continent – M. Woodcock-Collins

Objectives

To identify the external features of a bird.

To identify different types of birds by observing the external features.

To develop the skills of observation, recording, conservation and illustrating.

Activity

Go out and observe a few birds.

(1) Identify the different basic features of a bird, such as,

- beak : shape, size, colour.
- body : shape, size.
- neck : shape, colour.
- head : with or without crest.
- legs : webbed, long, thin, thick, short.
- wings : colour, shape.
- tail : long, short, rounded, forked, wedged.
- feather : colour, shape.
- eyes : lined, round, colour.

Draw these different features (separately) using simple diagrams.

Note the variations in colour, size and other external features and record them.

Draw the diagrams of the birds observed, using the form patterns.

(2) Listen to the Birdcalls

- Whistle : Bulbul, thrush, robin
- Chirps : Babblers, sparrows, finches
- Screech : Owls
- Croaks : Crows, herons, egrets
- Cackling shrieks : Woodpeckers, kingfishers
- Low moans : Pigeons, doves

Variation/Extension

By comparison the shapes of beaks and feet of different birds one can guess their feeding habits.

Birds have a variety of adaptations-including characteristics of beaks, feet, legs, wings and coloration. These adaptations have evolved so that the bird is better suited to its environment and lifestyle. A variety of major adaptations achieved during evolution are listed below:

Adaptation		*Bird*	*Advantage*
Beaks	1) Pouch-like	Pelican	Can hold fish, a food source
	2) Long thin	Avocet	Can probe shallow water and mud for insects, a food source.
	3) Pointed	Wood-pecker	Can break and probe bark of trees, for insects, a food source.
	4) Curved	Hawk	Can tear solid tissue, like meat, a food source
	5) Short, stout	Finches	Can crack seeds and nuts, a food source
	6) Slender, long	Humming bird	Can probe flowers for nectar, a food source

Adaptation		*Bird*	*Advantage*
Feet	1) Webbed	Duck	Aids in walking on mud, transportation
	2) Long toes	Crane, heron	Aids in walking on mud, transportation
	3) Clawed	Hawk, eagle	Can grasp food when hunting prey
	4) Grasping	Chicken	Aids in sitting on branches, roosting, protection

Identification of Birds :-

Helps in understanding the environment better.

For identification :-

1. Note down the size PLUs or MINUS a common bird (myna, sparrow, crow)
2. Distinguishing features (crest etc)
3. Prominent colours on wings, vent, rump, etc.
4. Shape and size of beak.
5. Call or flight pattern.
6. Nest.
7. Where found (habitat) : on the ground, in water, on the tree, etc.

Sr. No.	*Name of the bird*	*Size*	*Distinguishing features*	*Colour of body*	*Size and shape of break*
1.					
2.					
3.					
4.					
5.					
6.					
7.					
8.					

What these birds, which have decorated wings, beautiful body and artistic beaks, are thinking? Are they sending their message to mankind to listen to their preaching – only Action and Love could bestow the good and riches to you.

3.9 PANTHER'S APPEAL TO HUMAN TO PROTECT BIODIVERSITY

Dr. Ashis Kumar Ghosh

Panther, panther .... are you blight?
Yes sir, yes sir ....I am in a hopeless plight
Gone are the black-buck deers, gone is my home
The dense forest and valley where I used to roam

Gone are the native plants, sal, sagun and mangrove
Which were once my solituded treasure-trove

The gifts of nature which were once mine
Have now been usurped by your kith and kin

There is scarcity for food and pure water to keep alive
It now not known on which my offspring can survive
There is an outcry for my bones from the Beijing
For distilling of wines for the dollars they bring

They covet my ornamented skin, penis, jaws and claws
My nose, my teeth, even my faeces for good luck and imaginary cures

We are the aggressive in food-web of life
My extinction will also be the closing chapter of your life

Remember, don't wiped out, I am bravest and speediest
Lept your superstitions and obey your intellect

O *Homo sapiens* and O God Siva! When will you learn to refrain?
My death knell rings from behind the bamboo curtain

Try to preserve my cell in the bank of gene pool
It will give you variable progenies as a future tool

Six things come not back

The spoken word,
The sped arrow,
The time past,
The neglected opportunity,
The natural scenery and biodiversity lost,
The honour lost.

3.10 SACRED GROVE RELICS AS BIRD REFUGIA

Sacred groves are patches of natural vegetation dedicated to certain local deities, from which no harvesting of living matter is permitted by the local communities. (Ecologists currently recognize these scared groves (SG) as a unique cultural institution for conserving local biodiversity[1]). Numerous sacred groves are among the last representatives of climax vegetation in the Western Ghats and North East India, whereas SGs in other parts of the country have dwindled due to colonial land use policies (Gadgil and Subash Chandran 1992, Gadgil and Guha 1995) and erosion of traditional values regarding natural resource use. Nevertheless, hundreds of relics of SGs exist in the tribal tracts of Eastern India, and much of the tribal cultural life in West Bengal has been reported to be still centered around these relics (Deb and Malhotra 1997).

The observations we report here are a fallout of an ethnobiological survey we conducted from early April to June, 1996 in Jamboni, Jhargram, Gidhni, Belpahari and Banspahari Forest Ranges of western Midnapore district. In conformity with our previous study, we found remnants of SG in almost every tribal village in the region. Most of these groves, locally called Jahiristhan are relics of ancient SGs, containing 10-20 trees, amidst a denuded lateritic expanse. The tree species mostly found in these relics include sal (*Shorea robusta*), asan (*Terminalia tomentosa*), karam (*Adina cordifolia*), banyan (*Ficus benghalensis*), aswath (*F. religiosa*), pial (*Buchanania lanzan*), piya-sal (*Pterocarpus marsupium*), neem (*Azadirachta indica*) and mahua (*Madhuca indica*). All these species are also found in sal forests, albeit mostly in reproductively immature stages.

The sal coppice forest patches in the region under study are highly degraded and are now regenerating under community protection by villagers' Forest Protection Committees over the past 8-10 years (Deb and Malhotra 1993). The average crop height of the sal coppice stands is ca. 9 m., with GBH < 30 cm. However, the forest of Belpahari range is composed of a large population of old trees, with an average crop height of ca. 16 m. and GBH of 30-56 cm.

Although most of the SGs are composed of a score of old trees, the Kanak Durga temple grove at Chilkigarh is considerable large (ca. 20 acres), within the premises of the royal estate of Chilkigarh. This grove contains a large floral diversity and is characterized by old-growth trees, lianas and lichens. The canopy cover if this grove is atleast as good as that of the Belpahari forest.

The survey produced, *inter alia*, an inventory of local birds based on direct sighting records, for which we spent an average of 3 days in each of the five ranges during the period 2 April-30 June, 1996 and the duration of sighting effort was from 9 am till 4.30 pm each day. We have also incorporated here the record of bird sightings separately made in June-July 1995 in the same area. Although a number of waterfowl (e.g. pond heron, little cormorant, red wattled lapwing, stone curlew, little egret, cattle egret, white-breasted water-hen, etc) and migratory land birds (e.g. brown shrike) were recorded, we confine our discussion here to the resident land birds alone. A total of 42 species of resident land birds occur in the region studied. The birds were sighted in the sal forests, SGs, farm fields and vegetations in the settlements (Table 1). Since the original objective of the survey did not concern primarily with aviation diversity estimation in SGs, the records of bird occurrences are inadequate for statistical quantification and generalizations. Nevertheless, the data presented here may serve to indicate the apparent habitat preference of the birds with some overlaps. Table 3.10 shows that 22 land birds are found in the SGs, four of which are occasionally also sighted in the sal forest. The spotted dove was the only species that was found in all the three kinds of habitats. Ten species of birds were found only in sal forests and another 9 in both the sal forest and the settlement area, but not in SGs. A total of 19 birds were recorded to occur outside the sal forests; of these 14 were found in SGs and human habitations, and atleast four species (yellow-legged green pigeon, purple rumped sunbird, coppersmith and large Indian parakeet) were found only in the SGs. The house sparrow was found only in the human settlement area.

It seems plausible that the unavailability of sufficient quantum of fruits and the poor canopy structure in the regenerating sal forests fail to provide attractive foraging and nesting habitats for the birds occurring outside the sal forest. However, the Belpahari and Banspahari forests contain a good proportion of mature trees with well-developed canopy, yet only 3 out of the 21 birds recorded from SGs were sighted in the forests. (The socio-religious girdle of taboos around the SGs prohibiting trapping and hunting of animals seems to be the only explanation for the 18 birds occurring in the SGs, and not in the sal forests).

An important support to our conjecture that habitat protection is the primary reason for the bird's preference of the SGs, was lent by the local tribal hunters reporting that many birds that are hunted for meat (e.g. magpie robin, coppersmith, orioles, mynas and parakeets) tend to nest in Kanak Durga temple grove and other SGs. We couldn't verify these reportings, but were able to notice a small number of nests of the Indian myna and the black-headed oriole in three different SGs in Jhargram, Banspahari and Belpahari Ranges. An old (abandoned) nest of the weaverbird was also sighted on a palm tree in a derelict SG in Jamboni Range.

Further intensive studies were required to consolidate our findings. Nevertheless, assuming that the probability of sighting a bird species was equal in all three habitats categorized in Table 3.10, the record of 22 local land birds from the SGs suggests that these groves served as important sanctuaries for numerous biota in the past when they were of bigger physical dimensions. While it is unknown how many species of other groups of organisms were, or still are, protected by the institution of SG, this study indicates that these dwindling SGs continue to serve as important refugia amidst the landscape of species depletion. The function of SG in providing refuge to a variety of life forms has been appreciated by ecologists, but most studies in SG biodiversity have encompassed little more than floral inventories, thus leaving a considerable gap in the documentation of faunal diversity in SGs. Our qualitative data herein underscores the need for a rigorous study of the current ecological function of the existing SGs. Furthermore, this study reveals the efficacy of tradition in maintaining biodiversity in the face of continuing onslaught of the 'mainstream' economy on the indigenous cultural heritage.

Table 3.10 : Habitat preference of birds in West Midnapore

Taxa	*Common name*	*Sal forest*	*Sacred grove*	*Settlement area*
Apodiformes				
Apus affinis	House swift	+	-	+
Cypsiurus parvus	Palm swift	+	-	+
Caprimulgiformes				
Caprimulgus asiaticus	Common Indian nightjar	+	-	-
Columbiformes				
Columba livia	Blue rock pigeon	-	+	+
Streptopelia chinensis	Spotted dove	+	+	+
S. decaocto	Ring dove	+	+	-
Treron phoenicoptera	Yellow legged green pigeon	-	+	-
Coraciformese				
Ceryle rudis	Pied kingfisher	-	+	+
Coracius benghalensis	Indian roller	+	-	+
Halcyon smyrnensis	White breasted kingfisher (State bird)	-	+	+
Upupa epops	Hoopoe	+	-	-
Cuculiformes				
Centropus sinensis	Crow pheasant	+	-	-
Cuculus variuos	Common hawk cuckoo	+	+	-
Eudynams scolopacea	Koel	-	+	+
Falconiformes				
Milvus migrans	Pariah kite	+	-	+
Gyps bengalensis	Indian white-backed vulture	+	-	+
Galliformes				
Francolinus pondicerianus	Grey patridge	+	-	-
F. francolinus	Black patridge	+	-	-
Gruiformes				
Turnix susciator	Common bustard quail	+	-	-
Passcriformes				
Acridotheres tristis	Indian myna (Shalik)	-	+	+
A. fuscus	Jungle myna	+	+	-
Anthus novaeseelandiae	Paddyfield pipit	+	-	+

contd...

Table 3.10 – *contd...*

Taxa	*Common name*	*Sal forest*	*Sacred grove*	*Settlement area*
Cisticola juncidis	Streaked fantailed warbler	+	-	+
Copsychus saularis	Magpie robin (Doel)	-	+	+
Corvous splendens	Common crow	+	-	+
C. macrorhynchos	Jungle crow	+	-	-
Dendrocitta vagabunda	Indian tree pie	+	-	-
Dicrurus adsimilis	Black drongo	+	-	+
Nectarinia zeylonica	Purple rumped sunbird	-	+	-
Oriolous xanthornus	Black headed oriole	-	+	+
Orthotomus sutorius	Tailor bird (Tuntuni)	-	+	+
Passer domesticus	House sparrow	-	-	+
Ploceus philippinus	Weaver bird	-	+	+
Pycnonotus cafer	Red-vented bulbul	-	+	+
Sturnus contra	Pied myna (Go shalik)	-	+	+
Turdoides striatus	Jungle babbler	+	-	-
Strigiformes				
Athena brama	Spotted owlet	-	+	+
Tyto alba	Barn owl	-	+	+
Piciformes				
Dinopium benghalensis	Lesser golden-backed woodpecker	+	-	-
Megalaima haemacephala	Coppersmith (Basanta-bauri)	-	+	-
Psittaciformes				
Psittacula krameri	Rose-ringed parakeet	-	+	+
P. eupatria	Large Indian parakeet	-	+	-

* Included farm fields : + = Sighted;

3.11 Foot-print of Mammals

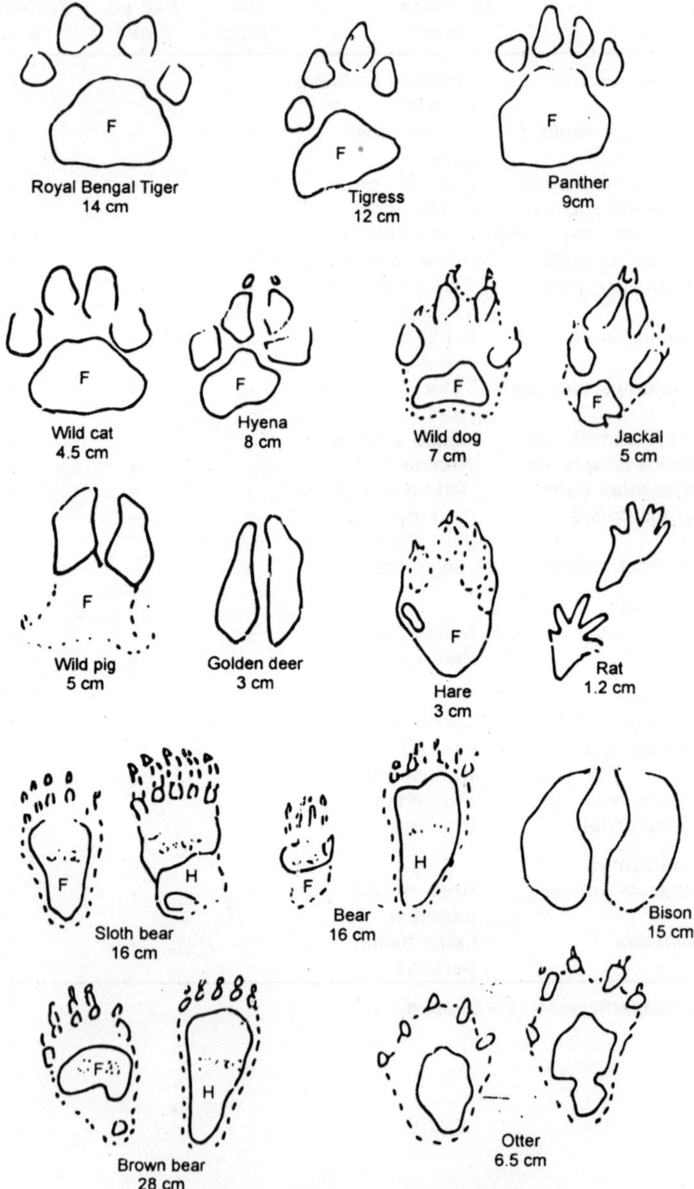

Right Fore Foot (F); Right Hind Foot (H)

3.12 APIPHILIC PLANTS IN AGRO-FORESTRY FOR BETTER BEE MANAGEMENT TO GET MORE HONEY AND WAX

Abstract

The integration among agriculturist-apiculturist, silviculturist-farmers and beekeepers has been adopted in Europe. But in India it is not found. Honey and wax, products of bees derived from beehives, has enormous curative properties. Honey production varies with the quality and quantity of nectar present in different flower species.

Introduction

Various uses of honey have been recommended from the Vedic era; such as priest in Panchamrita, physician in different ailments, dietician in various tasty foods, beautician in skin, etc. Besides wax is used in different industry. There are no diverse side effects of honey. So, honey is the best tropical medicine (Harris, 1994, Mutalik, 1991). Bees never like all the plants for their flowers as the quality of nectar varies from species to species (Suryanarayan 1986). They generally select the brightly scented flowers having sufficient nectar. To achieve particular quality of honey the nectar collection of the bees can be concentrated to a particular species of flowers by artificially spraying the flowers with 1% kerosene oil as the bees are too much sensitive. They prefer the smelling of 1% kerosene.

Materials and Methods

The mobile bee-keeping box may be placed in the needy selected field throughout the year in rotation according to the demand during the period of anthesis. A syrup will be made with the help of 100 gm sugar in 1 litre water and then 25 gm flowers will be immersed for overnight followed by a drop of kerosene oil to stimulate the bees at dawn. This flower's leachate is a very attractive drink to the bees. Now this syrup will be sprayed in the field to create a general affinity to that particular flowers. As a consequence the farmers as well as bee-keepers got together, their desirable honey and the crop/timber from the same field.

Mobility of bee-keeping box in rotation throughout the year is needed among different apiphilic plants community to get more honey.

Apiphilic Plants

1. *Brassica campestris,* Sarisha, Cruciferae. Herbs with yellow flowers having profuse honey. Flowering from December to February.
2. *Eucalyptus globulus, E. citrodora,* Myrtaceae. Tree having sufficient honey. Flowering from December to February.
3. *Acacia auriculiformis,* Akashmoni, Mimosaceae. Tree having yellow flowers and enriched in honey. Flowering from December to February.
4. *Shorea robusta,* Sal, Dipterocarpaceae. Deciduous tree. Flowering during February-March. Flowers whitish yellow.
5. *Mangifera indica,* Aam, Anacardiaceae. Evergreen tree. Flowering during February-March. Flowers yellowish white.
6. *Madhuca indica,* Mahua, Sapotaceae. Deciduous tree having abundant honey. Anthesis during February to April. Flowers pale yellow.
7. *Azadirachta indica,* Neem, Meliaceae. Deciduous tree having light yellowish flowers during April-May.
8. *Arachis hypogaea,* Chinabadam, Leguminosae. Herb having 2 types yellowish flowers during May-June.
9. *Avicennia alba,* Dulia baen, Avicenniaceae. Small tree. Flowers small, yellow. Flowering from June-August. Flowers are a rich source of honey and bee wax.
10. *A. marina,* Sada baen, Avicenniaceae. Shrub or small bushy tree, flowers pale yellow. Flowering from April-August. Flowers an excellent source of honey and bee wax.
11. *A. officinalis,* Kalo baen, Avicenniaceae. Medium tree, flowers yellow, small. Flowering from June-August. The flowers are a rich source of honey.
12. *Aegiceras corniculatum,* Khalsi, Myrsinaceae. Small tree. Flowers white. Flowering from April-September. The species forms a potential source for high quality honey and bee wax from its flowers.

13. *Sesamun indicum*, Til, Pedaliaceae. Herbs. White flowers blooming during May-June having sufficient nectar. The high quality honey derived from this plant have multiple curative properties.
14. *Nymphaea rubra*/Nelumbo nucifera, Padma. Nymphaeaceae. Aquatic herb having large flowers during June to October. High quality honey accumulates in these plants.
15. *Ceriops decandra*, Goran, Rhizophoraceae. Shrubs or small trees. Flowers white. Flowering almost throughout the year. Its flowers are a chief source of good quality honey.
16. *C. tagal*, Matagoran, Rhizophoraceae. Small tree, flowers white. Anthesis during March-August. Flowers are a rich source of honey and bee wax.

Conclusion

Adaptive trial should be undertaken under different agroforestry apiculture systems and the success and economy could be popularized among the farmers through the package of practices. Bees generally like the yellow and white flowers.

3.13 CLAY DYES OF PURBA MEDINIPUR – A GIFT OF NATURE

Abstract

The author has discussed the present situation of Potters engaged with potential natural clay-dyes with the view of a Social Scientist.

Introduction

Banak clay dye is an excellent natural dye obtained from under ground soil has been in use since 250 years. The Terra Cotta of Bishnupur (Bankura) and the clay dye of Purba Medinipur is well known to all. Clay dyes used in tiles, dolls, earthen pitcher and cooking pots, bricks, walls, etc. Age of the clay is 65 million years approximately. In South Bengal the colour of the clay varies extensively from white to brown to pink or black. But the Banak clay occurs at a depth from 7-8 ft of the crop fields. Generally the grains of clay are fine, medium sized grains are occasionally encountered. The Banak clay is mostly found in different villages under Tamluk, Panskura and Moyna Thanas. Some of the villages are Anantapur, Rajnarayan-chak, Dobandi, Maity-chak, Chakgarupota, Maguri, Tulya, etc. Here near about thousand families are attached with clay-dye industry. These families are familiar all over the state and even in India for the quality of their clay dye. Banak clay yields a very lusterous showy red dye which can protect from salinity.

Here the author has described the process and gradual development of clay dye potters from the very beginning. Potters usually make clay dye which are exported to the different districts of West Bengal, states of India and abroad.

Materials and Methods

Generally Banak clays collection periods are the summer months. Cost paid by the collector for clay dyes ranges from Rs. 4000-5000/- to the owner of a 720 square feet cropland. Usually 500 kgs Banak clay yields 50 kgs dye whose market value is Rupees one hundred and fifty only. Banak clay is first put into a series of covered earthen vessels alongwith sufficient amount of demineralised (Rain) water for 3 days. Here clays are processed for dye production by mechanical stirring through legs and then followed the precipitation techniques. Then in the second phase the upper supernatant solutions are placed in a series of covered earthenware pots in the exposed sunlight drying for 7 days. In the third phase the water alongwith soil mixture transferred into a series of covered earthen pots in the exposed sunlight drying for 90 days. A black colour soil cake is deposited in the inner wall of the pots which yields the valuable dye. The self life of the dye is quite stable in the temperature ranges from -4°C to 20°C. For this industry it needs minimum 4000-5000 linearly arranged earthenware pots. In Medinipur through "Rain Water Harvesting System" in large tanks rainwater has been collected round the year (88,000 litre/100 metre^2 both for dyeing and drinking).

Conclusion

Recently potters are facing various socio-economic problems particularly regarding finance, lands and markets. They are very much in distress because of the increased demand of asbestos, tin, polymer, etc. Potters are the pioneers of "Rain Water Harvesting

System". Government can implement through environmental rules that large multi-storied buildings are bound to made "Roof top Rain Water Harvesting System".

3.14 SOME VALUABLE ASPECTS ON BIO-NATURAL CULTIVATION

Abstract

Due to increased demand for ancient bio-vedic and bio-natural processes the author has tried to unveil some ecofriendly cultivation process.

Introduction

Effects of destruction of ecological balance by indiscriminate utilization of inorganic fertilizers, insecticides, pesticides, fungicides causes hazardous to living beings, scarcity of local ecofriendly animals like earthworm, insect, small fishes, arthropods, snakes and reptiles, birds, mammals, etc. which are usually seen in the field with their beneficial activities (through promoting fertility, pollination, pollution monitoring etc.) has been now gradually making a past historic story. So an alternative approach using bio-natural cultivation has been developed. After analysis of the causes and effects of the problem, awareness amongst the farmers has been made by the process. Emphasis has been made upon applying. Thus the concept to preserve the nature's ecosystem has been established. Foliar application of leaf extract on the plant's growth was considerably enhanced while disease incidence was markedly reduced by biocides and bio-fertilizers.

Text

Beneficial effect of Bio-natural process in the following ways :-

1. Longevity of the garden flowers can be enhanced.
2. Taste of different vegetable like small gourd, ladies finger, cow pea, etc. achieved the best.
3. Post harvest storage duration period increased.
4. Underground crop-yield can be increased by utilization of biopotash in the soil which can be obtained from ashes through burning of the dry abscised leaves of banana.

5. Mustard-aphids controlled by foliar spray of flour solution (8 gm flour in 1 litre water).
6. Swarming caterpillar controlled by foliaceous spray of the mixture of toddy containing *Datura metel* seeds dust.
7. *Lantana camara* and *Cajanus cajan* leaf extract act as a preventive measure against late blight of potato. Neem leaves also accredited the same result.
8. Foliaceous spray of water hyacinth leaf extract accelerates luxuriant growth of brinjal, potato, tomato etc., due to presence of gibberellin.
9. Foliax spray of *Dryopteris filix-mas* leaf extract increased resistance power against powdery mildew of pea.

Conclusion

The social thinkers and the scientists of this age tried to save the bio-diversity by changing the present day undesirable method of cultivation through biocides and bio-fertilizers.

3.15 PEST MANAGEMENT PRACTICES THROUGH NATURAL RESOURCES

Abstract

The article unveil the farmer's technological knowledge about pest management through natural resources which are cheap, easily available and having no side effect. These resources are fermented cattle urine, tobacco-neem-chilli leachate water, ash, kerosine or diesel etc. In Hinduism penned cows urine are sacred from the time immemorial.

Keywords: Pest, natural resources, cow urine, ash.

Introduction

Looking at farming till now more than hundred countries are solely depend on natural resources for disease and pest management. In this aspect they have already been developed as well as evolved some natural technological knowledge. Remarkable achievements found in African countries, like Ethiopia etc. To get rid off from pest and disease there are so many natural resources like wood ash, clay, whey, sand, castor oil, turmeric powder etc. There is a growing demand for natural resources because the

Table 3.15 : Natural resources alongwith the methods

Natural resources	*Name of crops*	*Name of pest*		*How used*
Wood Ash	House hold pulses and cereals, vegetables	Storage pests and other insects	-	Mixing with storage grains.
Chillies	-do-	-do-	-	Mixing with broken pods.
Salt	-do-	-do-	-	Mixing with storage grains.
Sand	-do-	-do-	-	Mixing with storage grains and specially in onion.
Cattle urine	Wheat	Aphids	-	Spraying the fermented urine in crop field along with water in (1:5) ratio.
Soap solution or detergents	Field pea	-do-	-	Spray in the field.
Tobacco (*Nicotiana tabacum*)	Cereals and pulses	-do-	-	Spray cured leaf-dust alongwith water (one leaf in 500 ml. water).
Marigold (Genda) *Tagetes erecta*	-do-	Storage pests	-	Dust of leaves and flowers.
Datura stramonium	-do-	-do-	-	Seed dust.
Lantana camara	-do-	-do-	-	Leaf dust.
Clay	Chick Pea	Beetles	-	Coating the seeds with clay
Kerosine	Wheat	Termites	-	Soaking the seeds for an hour before sowing
Whey	Pulses	Wilt	-	Soaking the seeds for overnight and then drying
Mustard oil	Chick pea, Musur (Lentil)	-do-	-	-do-
Neem powder	Pulses, vegetables, cereals	Aphids, flies, caterpillars	-	(25 gm in 300 ml water) Spraying the field/crops and storage grains.
Garlic powder	Stored cereals and pulses	Storage pests	-	Seed dressing
Turmeric powder	-do-	-do-		-do-

synthetic chemical pesticides are costly as well as causing pollution.

Materials and Methods

The study was carried out during field survey in different districts in remote areas round the year for three consecutive years. The survey was conducted by interviewing old aged farmers. The desired materials were soaked over night with sufficient quantity of fresh water. Cows urine are fermented for 7-10 days under direct sunlight. Spraying were repeated at 10-12 days intervals for each trial.

Results and Discussion

Search for solution to combat pests and diseases are mentioned here. Every man is innovative in nature, so they can expose their difficulties and limitation. Naturally they select locally available materials like ash, clay, sand, tobacco, sisal, aloe, neem, eucalyptus, castor, cow urine-dung, chilli, garlic, pyrethrum, nishinda etc. These cheap and revalidated natural resourses either used alone or in combination as a mixture are very effective. Ash, sand, chilli, salt, nishinda, turmeric, urine, dung were widely used for storage pest, although neem is often reported to be effective against wide range of pests and diseases by many scientists (Gautam 1994, Sreedevi 1994).

These safe socially acceptable and effective sustainable crop pest management for the developing countries needs special care because they are gradually extinct. These eco-friendly techniques evolved through farmers be financially encouraged.

3.16 NON-CONVENTIONAL OR ADDITIONAL AS WELL AS NATURAL SOURCE OF DIFFERENT ECONOMIC PRODUCTS FROM PALMS IN SOUTH BENGAL

Abstract

The author has tried to unveil the history, current status, prospectus and problems of palm industry alongwith the artisians though palm are "State Tree" of Tamil Nadu, still abundantly exist in different blocks and villages of South Bengal (such as Khejuri, Patna, Ajoya, Taldangra, Salda etc.)

Inroduction

Neera, palm gur, candy are famous during winter for their characteristic aroma, flavour and taste since the time inmemorial. The indigenous poor artisians of West Bengal are engaged in this industry and are tapping date palm (*Phoenix dactylifera*) and palmyra palm (*Borassus flabellifer*) to get sweet neera which ultimately yield gur, sugar candy and valuable nutritous-healthy drinks. Another neera and sugar yielding palm (water coconut) which is over exploited is *Nypa fruticans* (Golpati, palmae).

Text

The name "Khejuri" at present under Purba Medinipur derived for the plenty natural availability of Khajur gach "(date palm) and the Taldangara" under the Bankura district derived for the enrichment and luxurient growth of "Tal gach" (palmyra palm). These belts involves mainly with neera and thereby proper monitoring and utilisation of naturally grown wild palm trees has been occurred ago, now declined. So, some efforts to be taken by the Govt. of West Bengal to revive and propagate the indigenous palm trees in waste and barren lands. Palm gur making was an endangered job partiularly to the S.C and the tribals like santals, lodhas, bhumijs etc. As an alternate source of employment to the displaced tappers who use to make a toddy (country liquor) by farmenting sweet sap (neera) from palms. Still there is no registered palm Gur Co-operative societies in South Bengal.

There is no training of artisians in tapping preservation of neera, distribution for improved tools, equipment and arranging exhibitions and demonstrations etc. for its orientation on scientific and viable lines for the industry and upliftment of tapper artisians (Kamble 2003). It is a winter season based industry providing employment on an average of 150-160 days in a year. There is no conducting research and development of new area of production to utilise abundant and valuable natural resources i.e. palm trees which are otherwise un-exploided. Popularising products of neera as beverage as well as its nutritive value is well recognised. Though it is a tradition industry for the tapper community (as their main seasonal occupation for livelihood in the state). Still no other community will accept tapping of palms as an additional source of income due to its following hazardous nature of work:-

(a) Tree climbing process for tapping trees is a most difficult task at down and on afternoon.

(b) Youth S.C and Tribals does not find tapping occupation and is not adequately remunerative, having storage problems, risky marketing problems due to cut-throat and keen competition.

Conclusion

The existence of yet numerous untapped vast potential trees for further development of this idustry on a wider scale for offering employment to a large numbers of migratory artisians. The new implemented devise by M/s Mary's Industry in automatic climbing of trees should be distributed among the artisians. For reducing the risk for climbers aerial rope ways and bamboo ways have been introduced. Concrete scientific method should be developed for preservation of neera, palm gur and other related products. Regular, irrigation can increase the exudation. A very efficient furnace through biogas or solar energy having multiple pan system have restore the fuel cost. Neera have an anti-diarrhoeal property.

3.17 ETHNO-NUTRITIOUS SELF-MADE BABY FOOD FROM PASCHIM MEDINIPUR

Abstract

The author has tried to focus the natural life-style of the poor villagers along with their baby through naturally available household nutritious self made baby food "Sneha".

Introduction

"Sneha" is well known for its useful porperties for babies sound body. It is healthful because it is enriched with different nutritious components which is essential for sound growth of 1-6 years baby as sound body keeps a sound mind. This ethnic food is used on the location specific demand, culture, economy, food habit and needs (Dutta and Dutta 2005).

Survey Method

Survey was conducted during 01-01-2002 to 05-01-2002 among differet women is Amrakuchi Gram Panchayat and the author gathered valuable information from Sephali Dolai, Padma Dolai, Bandana Bhuina, Chandana Bhuina, Piyali Bibi etc. during his investigation and survey.

Material and Methods

Washed good quality of wheat (*Triticum aestivum*) and moong (*Phaseolus aureus*) pulse in fresh water and then dried at direct sunlight. Then fried separately and mixed in (4:1 ratio) followed by grinding and pouched for marketing. There are 19,000 self-supporting females group are present in Paschim Medinipur.

Conclusion

This food have immuno-strengthening properties and encourage the growth of babies. The baby food 'Sneha' is far better then any other artificial food making company which exists and produces baby food like amul, amulya, horlicks etc. as it is cheaply available and natural food having no side effect.

Nutritive value of 'Sneha'

Calory	1732 Kcal
Fibre	5.6 gm
Protein	701.7 gm
Carbohydrate	344.7 gm
Fat	7.2 gm
Folic acid	286 gm
Phosphorus	1629 mg
Thiamine (B_1)	2.27 mg
Riboflavin-(B_2)	0.89 mg
Neacin-(B_5)	24.4 mg
Carotene	305 :-30 x 5 micro gm
Fe	28.1 mg
Ca	249 mg

The processing and preparations are the treasures of food heritage of tribal women.

3.18 THE NATURAL METHOD OF HEALING OF ETHNIC PEOPLE THROUGH SOIL THERAPY AND HYDROTHERAPY

Abstract

The practice of natural therapeutics is curative for many common ailments in every day life. The purpose of this paper is to explore scientific creditability of narration of soil and water in the scriptures of the world and traditional usage the biological activities of soil and water on human body is very significant. It is emphasized that these naturally available soil and water are having high therapeutic value and must be consumed whenever necessary as a household remedy.

Keywords: Hydrotherapy, Hydrotuberation, Compress, Gargling.

Introduction

Civilised man is the sickest among the animals in the world as they far from the nature. So, they are often suffer by different ailment. Due to rapid diminishes in natures gift and gradual increases in cost of life saving synthetic medicine, there is a vast demand on natural method of healing by air, water and clay, be your own doctor should be our slogun. It is better to wear out, than rush out. If we fail to summon sleep at will, there must be something wrong with our method of auto-suggestion. As medicines undertake to cure one disease by producing another (side effect present). Practically all diseases are constitutional. Disease is a natural process and its symptoms are the reactions of the body to the disease. So, the chief function of the physician is to aid the natural forces of the body.

To keep the body fit the following general rules should be followed:-

1. Day by day in every way I am getting better and better.
2. Laugh and grow fat.
3. Keep away from "hurry and worry".
4. The time spent in rest is an investment for the future. So rest absolute rest is the panacea. Rest permits accumulation of waste to be excreted and the destroyed material replaces. Bed rest often suffices to relieve all the sympoms in mild cases.
5. Practically all diseases are aggravated by lack of sufficient oxyginated air. Fresh oxyginated air at dawn is a nerve stimulant.
6. Constipation is the root cause of all diseases of civilisation due to intake of fast food through auto-intoxication.
7. No one ought to take cold bath unless completely warm.

8. Instead of using medicine, rather fast a day with lemon juice.
9. Intake of fresh fruits and vegetables daily is the most natural healing remedy.
10. Before bath skin-rubbing exercise by mustard oil in open air treatment for mental order. So long as the sun feels good, it will do you good. If the sunbeams fall upon the naked skin then it will have many curative property.
11. More people die from eating too much than from eating too little. Less food means more rest for the inflamed intestine.
12. Suffering becomes beautiful when any one bears great calamities with cheerfulness.
13. Balanced diet cure more than doctor. That what is prone to cure, is prone to kill.

Text

Nature alone (air, water, clay, exercise etc) can cure, many a local trouble will free from an earth compress and water compress as if by magic, such as:

1. **Sandy soil:** Lukewarm sandy soil bag used in swelling of neck and in congestive headache. Washed the teeth by sandy soil as a tooth paste to prevent and cure toothache and swelling of gum.
2. **Loamy soil:** Cold earth compress used in stomach inflammation, diarrhoea, cholera, constipation, piles, gastric and duodenal ulcer. Cold earth compress upon belly will be changed at one hour interval. Besides cold earth compress used in migraines, boils, abscess, carbuncles, cut and wounds.
3. **Saline soil:** Intake of 1gm. soil used to prevent thyroid problem and in colic pain. In lice infestation rubbed the soil in hair before bath and then washed out.
4. **Clay soil:** Rubbing the skin with liquid clay in skin disease and in dandruff which act like a soap.
5. **Lime soil:** Intake of 2 gm soil daily as a preventive measure against worm infection. In sprain heating earth compress alongwith turmeric and a pinch of edible salt is effective.
6. **Red Soil:** Intake of 2 gm soil as a daily dose in anaemia.

7. **Ash:** Rubbed the ash in skin to prevent from gout, cold and cough. It act as an antiseptic in cut and wounds. And also used as a tooth paste.
8. In Bastar region people use soil on which Safed Musli grown. This soil is collected when the life cycle of Musli herb is over. They use the soil in treatment of many common diseases (mostly externally), even in the treatment of cancers. The soil collected under the tip of Musli root is considered the best.

Hydrotherapy

1. Alternate hip bath cures dyspepsia, missed abortion, indigestion.
2. Hot soot bath an effective measure against cold and cough.
3. Sitz bath for 20 minutes daily cure impotence, insomnia, neurosis.
4. Cold compress is curative in fever, sprain and burns.
5. Fomentation on stomach by hot water bag after meal for 1 hour have a precaution against dyspepsia and indigestion.
6. Alternate compress by hot and cold water in respective organ is very effective in bed sore, inflammation, muscle paralysis, tumour, intoxication, sprain etc. as no bacteria can alive. Similarly chest pack, throat pack, foot pack can be done.
7. In sea sickness, tetanus and histiria ice pack in vertebral column is curative. Ice pack also used in sprain and burns.
8. Hydrotuberation in uterus promotes fertilization.
9. Rubbing the bowels with coconut oil alongwith water as a cold compress cures colic pain and dysentery.
10. Alternate gargling with a pinch of edible salt in affected throat due to cold is a step for natural healing.
11. Drinking of sufficient cold water and lukewarm water promotes digestion, prevents constipation, gastric troubles.

Conclusion

Earth and water are rich potential source for human and they probably serve as chemical defensive against infection. Instead of using medicine these natural resources (earth and water) are enough for maintaining our health. The art of healing is inexpensive and at the same time scientific which may be recommended to every home.

3.19 NATURAL METHOD OF HEALING OF ETHNIC GROUPS THROUGH COMMON PRACTICES AND DIETOTHERAPY

Abstract

Though ethnic people are painstaking worker still they are well acquainted with dietotherapy from the time immemorial. As they are poor they are less dependent on costly modern medicines. Their life style and food habits keep them strong and active. Intake of fresh natural instant seasonal fruits, food, drinks and their method of simple life style may be the alternative of modern medicines. They belief diet is health, diet is medicine.

Keywords : Ethnic group, dietotherapy, laxative, household.

Introduction

The demand of dietotherapy gradually increases as the allopathic and homeopathic medicine have some harmful side effect. So, the chief function of the modern physician is to treat the patient by dietotherapy and aware the art of living. It is proved that appropriate recommendation about food-drink and art of living is the best remedy for many ailments. Eat your way to health and avoid vitamin capsules and tonic. Attempts at correction should be directed towards the diet.

Method of Survey

The study is based on frequent household surveys of different localities of Bankura, Medinipur and Purulia districts. Formal and informal interviews were conducted with regards by applying some of the adopted techniques.

Text-Result and Discussion

The author enlightened the following effective natural method of healing:-

1. In England there is a proverb that "for every child a teeth". So, at the late stage of pregnancy and during lactation

period to meet the requirements of calcium and phosphorus til (*Sesamum indicum*), bakphul (*Sesbania grandiflora*) flowers, noteyshak (*Amaranthus tricolor*), small fishes and whey should be eaten. Besides in the initial stage of pregnancy neera, sugarcane juice, Jaggery, molasses and honey is effective to prevent vomiting alongwith small quantity of husking pedal-rice, coarse flour, boiled coarse flour of wheat and parched paddy.

2. Heavy diet diminish the secretion of stomach juice as a consequence developed hyperacidity, flatulence, piles, diabetes, gout, blood pressure, liver and gall bladder problems.
3. Hunger is the best sauce. So, always food intake will be in extreme hunger. Food intake with gladness is the best process for easy digestion rather than shocked, hurried, worried, depressed and stressed condition. After meal napping promotes digestion. Juicy fruits can be eaten at any time during the day though Indian monks intake fruits at evening. Refined and concentrated fast food causes constipation due to lack of peristalsis. It should be remember that sprouted potato and sweet potato produces toxic substances (Henry Edward Cox, 1950). Use of tamak promote senescence. If tamak is given in the tongue of a hare or on the naked skin of a rat induces death.
4. It should be remember that fever is nothing but a natural process for elimination of toxic substances from the body. Lukewarm water should be drunk during ague only. As hot water causes fatigue but cold water is energetic.
5. Excessive tea causes dyspepsia, insomnia, loss of appetite, heart problems. Addition of milk in tea promotes its quality. Constipation causes by tea, coffee, Coco, Chocolates, Joan, flattened rice puree, karpur, fried food, fast food, Kanchkala (*Musa paradisiaca*), flour, arrowroot biscuits, milled rice, sugar, gandal (*Paederia foetida*) mauri, amlaki, ghee. Constipation causes accumulation of uric acid thereby producing rheumatism, gout, arthritis. Here lemon juice is effective. Fish, meat and egg produces uric acid. So, in this condition salads and sour fruits is a complimentary food.

6. Laxatives are chutneys of tamarind with molasses, ripe, banana with a pinch of salt, burning bel (wood apple) alongwith molasses, whey, cream.
7. To cure acidosis musk-melon, orange, raw potato, coconut with fried rice, lukewarm water for instant relief. Though sodium bi carbonate may be fed but it is harmful. Toasted and roasted loaf, ground nut oil and coconut oil can be fed as are easily digestible.
8. For diabetes inevitable foods are letuce (*Lactuca sativa*), coconut, soybean, curd, whey, sufficient water and often starvation.
9. In blood sugar cake, condensed milk, canned fruit in syrup, cookies, chocolates, chewing gum, doughnuts, honey, jam pies and pastries, pudding, soft drinks, sugar, sugar coated cereals should be avoided.
10. All kinds of vegetables can be eaten raw in the form of salad. Sweet potato is more beneficial than potato as it is enriched with vitamin A. During the preparation of curry a pinch of salt should be added at first to escape from losses of vitamins and minerals. Heavy use of paddle during cooking promotes vitamin loss in often close contact of oxygen. Fast cooking causes less loss of vitamin. Use of sodium-bi-carbonate destroy vitamin C, thiamine and riboflavin (which defer senescence). Reheating of fomenting milk, cooked food and curry causes loss of vitamins. Fried egg and foods are not easily digestive. Excessive use of spices alongwith constiveness or tightening the curries induces weaken stomach thereby producing many liver-heart nerves problem.
11. Always body weight should be slightly below the normal at middly age (40 to onwards) which enhances life span (USA, Dept of Agriculture, Food and Life, P. 121, 1933). Under weight and starvation is the best tonic for high blood pressure.
12. In heart problem eating should be saltless food so that the blood stream remains pure. Excessive fat deposition increased the labour of heart thereby reducing the activity of muscles and skin. Butter, cream, cheese, ghee, fried food

greasy food, junk food, red meat (specially those with visible fat) cause heart problem and obesity, obesity also causes high blood pressure, diabetes, heart diseases. So, the diet should be the juicy fruits, vegetables curries.

To prevent anaemia mamals' liver and kidney is the best. Both half-boiled and raw liver alongwith tomato juice is effective (if fed) for easy blood preparation. Iron easily absorbed in the body when iron intake will be alongwith copper.

To increase life span or longivity curd is beneficial than cheese. There is a myth among ethnic people to extend life span the power of milk products are as follows:-whey>curd>cheese>milk. Fat is avoided during constipatian, dyspepsia, diarrhoea, colites, gout, high blood pressure, heart problems and skin disease.

Skimmed milk (whey) is a calcium, phosphorus, iron riched but fat free cheap food for the poor. So it is best for heart and to defer senescence as it has high nutritional status which is essential for increase longivity. During the preparation of curd by adding a pinch of whey, the quality of curd is improved. All the sunny cows in herd i.e. which are in close contact of sun and nature their milk and whey are enriched with vitamin-D. Whey is very effective to kill all the harmful bacteria of human intestine. So, butter milk is of inestimable in most of these cases. The *Lactobacillus acidophilus* in curd kills many germs of intestine, produce vitamin-B, and defer senescence. In scarcity of breast milk always fed goat's milk. Because goat's milk is best for babies to keep away from all the diseases. Curd also keep away the babies from diarrhoea and droppings problem but barley causes dyspepsia. Honey is a predigestive food and diuretic whereas egg-containing fishes are – having low calories and toxic.

Some Interesting Facts About the Human

1. Babies are born without knee caps.
2. Relative to size the strongest, muscle in the body is the tongue.
3. Children grow faster in the spring.
4. Women blind nearly twice as much as men.
5. If you go blind in one eye you only lose about one fifth of your vision.

Table 3.19 : Important food used as medicine in natural method of healing by Tribals

Name of ailments	*Local Name*	*Botanical Name*	*Parts used*	*How used*
Nephritis, mental stress, loss of apetite, high blood pressure, obesity	Kagzi	*Citrus aurantifolia* (Rutaceae)	*Fruit*	Drunk the juice alongwith water in empty bowel.
Obesity				Drunk the leafy vegetable juice as a under weight tonic
Cholera	Bel	*Aegle marmelos* (Rutaceae)	Fruit	Recommended as a medicated drink.
Paralysis	Hing	*Ferula asafoetida* (Umbelliferae)	Seed	5 gm. paste rubbed daily as ointment.
Thinness	Ground nut	*Arachis hypogaea* (Leguminosae)	Cotyleden	Fed it alongwith banana and fresh curd though banana is the chief food in middle Africa.
Eczema, histiria, measles, pox, high blood pressure, nerve problem				Saltless food and curries hasten cure these diseases.
Insomnia, nerve disorder	Onion	*Allium cepa* (Liliaceae)	Bulb	Fried onion promotes these.
Missed abortion, goitre, deformed child				Regular intake of sea animal, is recommended particularly in hilly area due to lack of iodine in soil.
Missed abortion, deformed child, asthma, heart problem, obesity	Germinated wheat, Wheat bran oil, coarse flour	*Triticum aestivum* (Gramineae)		Fed these alongwith molasses.
	Letush	*Lactuca sativa* (Compositae)		
	Spinach (Palong)	*Spinacea oleracea* (Chenopodiaceae)		

6. The length of the finger dictate how fast the fingernail grows. Therefore, the nail on your middle finger grows the fastest, and on average, toenails grows twice as slow as fingernails.
7. Odontophobia is the fear of teeth.
8. Hair is made from the same substance as fingernails.
9. Human blood travels 60,000 miles per day on its journey through the day.
10. An eyelash lives about 150 days before it falls out. One in every 12 men is colour blind.
11. About 400 gallons of blood flow through the kidney/day.
12. Placed end to end the blood vessels would measure about 62000 miles and the average human brain has about 100 billion nerve cells.
13. During sneezing all bodily functions stop even the heart.

Conclusion

From the above discussion the contribution made by the ethnic people through natural method of healing to modern system of medicine is worth nothing. The efficiency has been established by results from different persons.

3.20 APIPHILIC PLANTS IN AGRO-FORESTRY FOR BEE MANAGEMENT

Abstract

Bee hives yields valuable multiple products such as royal jelly, wax and honey which are closely associated with the daily life for medicine and cosmetic etc. No product made by man can mimic royal jelly, bee wax and honey made by bees. The author has tried to focus the apiphilic plants to get more bee hive product round the year through migratory transit camps.

Introduction

Generally the mangrove swamps of Sunderban is famous for its bee collectors (mouli) and honey as it is enriched with nectar yielding plants like dulia baen, sada baen, kalo baen, khalsi, goran, matgoran, tora, keora (*Sonneratia* sps.) etc., which produce nectar more on less throughout the year. Migratory honey collectors start their journey towards Sundarban in search of honey from October and the transit camps end on May. Besides Sundarban, North 24 Pgs and Nadia districts are the Chief source of nectar in off-season (for 5 months) in West Bengal. So, naturally bee-keepers exhibit migratory habit and crowded in these regions followed by transit camps and boxes during off-seasons in different gardens, forests and crop fields, profuse nectar yielding plants are found heavily in these districts :

1. In North 24 Pgs and Nadia plenty occurrence of jam (*Syzygium cumini/jambos*), litchi (*Litchi chinensis*), kul (*Ziziphus mauritiana*), dhone (*Coriadrum sativum*), mustard (*Brassica campestris*).
2. Medinipur and Bankura famous for Eucalyptus.
3. Purulia enriched with kul.
4. Hoogly district having plenty of til (*Sesmum indicum*), litchi.
5. Malda, Murshidabad and Uttar Dinajpur are famous for am (*Mangifera indica*), Til, litchi, mustard, sun-flower.

Materials and Methods

Unemployed persons can start their business for self establishment with the financial assistance of co-operative or Govt. Bee hive containing boxes can be purchased in lieu of Rs. 4000/- per box having a queen (longivity 3 years), 500 drones (longivity 25 days) and 10,000 worker (longivity 3 months). Boxes are rectangular (10″ × 16½″ × 20″). Boxes will be placed in the garden/forest/crop field in 3 ft. height from the ground level to escape from toad, wasp etc.

Text

Colour, taste and smell of honey varies depending on the quality of nectar (Suryanarayan 1986). Such as litchi's honey is light green, keera and mustard's honey is white and most of the flower's honey is straw or light golden in colour. The most valuable product of beehives is royal jelly whose cost is Rs. 1,500/- per kg which is made up of pollen and bee's saliva. Honey yield/box/month is 10-15 kg whose value is Rs 160/-. Honey yield in 100 box/year whose cost is Rs 2 Lakhs (approx) and net profit will be Rs 75,000/-

(approx). In off seasons the expenditure for bee's food (sugar solution) is 2700 kg sugar for 100 boxes, value is Rs 40,000/-.

Conclusion

Bee-keeping and collecting should be the hobby for gardeners, cultivators, forester, students and unemployed persons because bee keeping not only gives bee products but also increased the yields of flower growers, orchards, forces and crop yield through inducing high rates of pollination. Honey yield depends on the availability of nectars and the quality of bees.

Instant Household Method for Honey Preservation

Honey is hydrophilic. Generally honey absorb water-vapour from the atmosphere and then fermented within a few days.

To avoid this event honey container deeped for few minutes in boiling water bath still smokes (vapour) arise from the honey followed by preservation in air tight glass-container.

Honey is too much sensitive to pungent smell (like petrol, cabbage, coal-tar, pine, metal-container etc). It absorb pungent smell from the container and also from its close circumference.

3.21 MEDICINAL PLANTS USED FOR TREATMENT OF DIABETES BY THE TRIBAL PEOPLE OF BANKURA, PURULIA AND MEDINIPUR OF WEST BENGAL

Abstract

Diabetes is the root cause of all types of disease and damages to kidneys, wounds, eyes, nerves and other organs. Diabetes and high blood pressure accelerates impotency, premature senescence. Important household unexplored different medicinal plants which are used to cure diabetes discussed here.

Keywords : Diabetes, Hypoglycemic, Insulin

Introduction

Every sixth diabetic in the world is an Indian. Diabetes is of two types Type-1 insulin dependent and found within the teenages. Worldwide insidious increase in type-2 diabetes or non-insulin dependent diabetes mellitus (NIDDM) is a matter of global concern,

90-95% diabetic patients are type-2 (Shawl et al, 2004). It's concern that need immediate action. Symptoms of diabetes are :

i Tingling or numbness in arms and legs.

ii Increased thirst, appetite, urination, apathy.

iii Unexplained weight loss.

iv Undue tiredness.

v Itching in sexual organs.

vi Non healing wounds, repeated infection and visual disturbness.

In addition to many modern medicines there are so many very effective traditional medicines available in public domain for the treatment of diabetes. There are more than 400 plants species showing hypoglycemic and may remain unexplored and to be scientifically evaluated (Natural Product Radiance, Vol-3, pp-185). There is an increased demand to use natural products with anti-diabetic activities due to the side effects associated with the use of insulin and oral hypoglycemic agents. Sugar and starches in food are broken down into glucose which then circulates in the blood. The hormone insulin take glucose to be used for energy or made into fat. But people with type-1 diabetes do not produce enough insulin. Those with type-2 diabetes produce it, but have lost sensitivity to it. Having too much glucose in the blood can cause serious long term damage to eyes, kidneys, nerves and other organs.

Bankura, Purulia and Medinipur are the tribal districts of West Bengal. These districts are mostly inhabited by Santhal, Oraon, Koras, Bhumij, Mech, Lodha, Mahali and Munda. These men in their own unique way apply herbal medcines to encounter various ailments. The indigenous system of medicines in these districts (i.e. West Rardh) has almost remained unexplored except for a few reports from Chowdhury et al, 1982, Pal and Jain, 1989, Ghosh and Das, 1999. Bankura and Purulia lying between 22° 46′-22° 38′ N and 86° 36′-87° 46′E parallel. Whereas Medinipur is in the South-western corner of the state (between 21° 36′ 40″ and 20° 56′ 40″ N and 86° 35′ 22″ and 88° 31′ 30″ E). Its areas are 30000 km^2. Forest cover about 4000 km^2. Here the Santals and Lodhas are the main trbals. Santals are about 1 Lakh distributed throughout these areas. Lodha populations are about 35000 confined mainly in northwestern part of this regions. Other tribal populations are within 10000 each.

Observation

It is an outcome of three years survey in tribal areas of Bankura, Purulia and Medinipur among knowledgeable folk-doctors as they posses inherited knowledge regarding the plants of ethnobotanical importance. The information furnished by them were recorded together with the collection of the plant specimen used by them. Vaucher specimens alongwith taxonomic determination made and deposited in the herbarium of botany department, (Ramananda college Bishnupur). The medicine men were interviewed at the spot using a questionnaire. As a strategy to confirm the information it was rechecked with other herbalists.

Discussion

Cinnamon lowered blood levels of fats and bad cholesterol, which are also partly controlled by insulin. It neutralized free radicals, damaging chemicals which are elevated in diabetes. Since blood sugar tends to increase with age accelerating aging by crosslinking with proteins (glycation), the ability of tea to lower serum glucose levels is extremely important as part of it anti-aging benefits. Tea inhibits both salivary and intestinal amylase so that starch is broken down more slowly, and the rise in serum glucose is thus minimized. Tea may reduce the intestinal absorption of glucose. A relatively little known compound found also in onion and specially in green tea called di-phenylamine seems to have a strong sugar lowering action (Karawya et al, 1984). Diabetes is linked to the bodies inability to control levels of glucose in the blood often because the hormone insulin is not functioning in a normal way. Treatment with the extract of different herbs was found to normalize blood sugar levels and to boost the way insulin works. The main food of Iyerland is potato but in India peoples avoid potato in diabetes. Newly harvested sproutless upto two months aged potato having low glucose can be recommended at low amount for diabetic diet. After two months potatos starch is gradually converted into soluble sugar. The ideal sweetener for diabetes is *Stevia rebaudiana* Bertoni. The ash of peacock tail 125 gm alongwith few drops of honey if taken daily is very effective in asthmatic-diabetic patient.

Table 3.21 : Medicinal Plants Used As Potent Antidiabetic Activity

Botanical name & family	*Local name*	*Parts used*	*Method of preparation and use of medicine*
Cinnamomum zeylanicum Bl (Lauraceae)	Daruchini	Bark of the stem	½ a teaspoon dust or even by soaking a stick in tea.
Camellia sinensis L. (Theaceae)	Cha	1st and 2nd leaf and the bud	Fed the mixture twice daily in empty stomach. – –
Allium cepa L (Liliaceae)	Piaj	Bulb	Used as raw vegetable alongwith rice.
Datura metel L. (Solanaceae)	Dhutra	Seeds	Crushed the seeds. Fed 25 mg/kg body weight once daily.
Rhizophora apiculata B1 (Rhizophoraceae)	Garjan	Leaves	Crushed the leaves. Drunk the extract 4g/kg body weight once daily.
Achras zapota Linn. (Sapotaceae)	Safeda	Roots	Soaked the root for overnight. Drunk the alcohol extract 20g/kg body weight.
Aegle marmelos Corr. (Rutaceae)	Bel	Leaves, Ripe fruits	Extract the leaf juice. 2 teaspoonful juice twice daily in empty stomach with few drops of honey.
Costus mexicanus DC. (Zingiberaceae)	Costus	Leaves	Ate one leaf daily.
Punica granatum Linn. (Punicaceae)	Dalim	Pericarp	Fed one fruit daily.
Mangifera indica Linn. (Anacardiaceae)	Am	Tender leaves	Oral administration of aqueous extract of the leaves 1g/kg body weight.
Syzygium jambos Alston (Myrtaceae)	Jam	Seed	Crushed the seeds. Ate one spoonful of seed dust twice daily.

contd...

Table 3.21 – *contd...*

Botanical name & family	*Local name*	*Parts used*	*Method of preparation and use of medicine*
Withania somnifera Dun (Solanaceae)	Aswagandha	Root	Drunk 3 gm root decoction once daily alongwith 3 ml leave extract of Thankuni (*Hydrocotyle asiatica* L.)
Trigonella corniculata Linn. (Papilionaceae)	Methi	Seed	Soaked the 5 gm seeds overnight. Drunk ½ cup leachate water once daily.
Azadirachta indica A. Juss (Meliaceae)	Neem	Leaf	Ate 5 gm leaf dust once daily alongwith 5 pepper.
Vinca rosea L. (Apocynaceae)	Nayantara	Leaf	Ate 3 leaves once daily at dawn.
Tinospora cordifolia (Menispermaceae)	Gulancha	Assimilated root	Soaked 2 inches fresh root in ½ cup cold water. Drunk the leachate water once daily.
Gossypium herbaceum L. (Malvaceae)	Kapas	Seed	Ate 2 gm raw seeds twice daily.
Acacia catechu Willd (Mimosaceae)	Khayer	Gum	Ate one gram alongwith betel leaf once daily.
Musa paradisiaca L. (Musaceae)	Kola	Plantain	Extract the juice. Drunk 2-3 teaspoonful juice twice daily alongwith 3 drops of honey.
Portulaca oleracea L. (Portulacaceae)	Nuney shak	Aerial part	Boiled 10 gm in 4 cup of water. Drunk the filtered juice twice daily.
Hibiscus esculentus L. (Malvaceae)	Dheros	Fruit	Two vertically dissected fresh fruits soaked overnight in ½ glass cold water. Drunk the leachate in every morning. It is very useful for diabetic and blood cancer patients.

contd...

Table 3.21 – *contd...*

Botanical name & family	*Local name*	*Parts used*	*Method of preparation and use of medicine*
Shorea robusta Gaertn (Dipterocarpaceae)	Sal	Tender leaf	Crushed 2 leaves in ½ cup cold water and drunk the juice once daily.
Hydrocarpas kurzii Warb (Flacourtiaceae)	Chalmugura	Seed	Fed 1 gm seed dust twice daily.
Cocos nucifera L. (Arecaceae)	Coconut	Kernel	Fed daily.
Lactuca sativa L. (Compositae)	Letuce	Leaf	Drunk 2 teaspoonful juice regularly.
Glycine max L. (Papilionaceae)	Soybean	Seed	Used as vegetable in curry.
Cuminum cyminum L. (Ranunculaceae)	Sajira	Seed	Fed 2 gm seed dust once daily.
Allium sativum L. (Liliaceae)	Rosun	Bulb	Fed a raw bulblet once daily.
Emblica officinalis Gaertn (Euphorbiaceae)	Amlaki	Fruit	3-4 gm dry rind ate daily alongwith a pinch of rock salt.
Cajanus cajan L. (Papilionaceae)	Arhar	Leaf, root and seed	Drunk 2 teaspoonful juice once daily with few drops of honey.
Diospyros exsculpta (Ebenaceae)	Kend	Stem bark	Drunk 5 ml juice. Extract act as hypoglycemic.
Terminalia catappa L. (Combretaceae)	Deshi badam	Fruit	50 gm fruits is sufficient for daily dose.

contd...

Table 3.21 – *contd...*

Botanical name & family	*Local name*	*Parts used*	*Method of preparation and use of medicine*
Ficus benghalensis L. (Moraceae)	Bot	Bark	5 gm bark soaked overnight. Leachate is very effective against diabetes and cholesterol.
Ipomoea digitata L. (Convolvulaceae)	Bhui-kumra	Root	Prominent role has been found if drunk ½ cup fresh root extract once daily.
Artocarpus heterophyllus Lamk (Moraceae)	Katahal	Leaf	Crushed the fresh tender leaves. Fed ½ cup juice once daily.
Cassia fistula L. (Caesalpiniaceae)	Bandarlathi	Pulp of the pod	Safely used for longer period as daily dose of 5 ml.
Cucumis sativus L. (Cucurbitaceae)	Shasha	Fruit	To reduce sugar ate a freshly collected green fruit after lunch once daily.
Santalum album L. (Santalaceae)	Swet Chandan	Stem	Fed 1 gm paste daily.
Momordica charantia L. (Cucurbitaceae)	Karela	Fruit	A burnt bitter gourd if eaten daily there will be no fear for diabetes.
Hollarrhena antidysenterica (Apocynaceae)	Kurchi	Seed	Soaked 2 gm seeds for overnight. Drunk ½ glass leachate cold water in empty stomach.
Terminalia arjuna W & A (Combretaceae)	Arjun	Stem bark	Extract used in biosweets for reducing blood sugar.
Daucus carota L. (Umbelliferae)	Gajar	Root	Extract used in biosweets for reducing blood sugar.
Moringa oleifera Lam (Moringaceae)	Sajina	Stem bark	Extracted juice is recommended thrice daily for 7 days.

Conclusion

Most of the recipes include only one plant species, however some preparations are the combinations of several herbs. More over, the same plant is used for more than one disease. So, it is difficult to say which plant is actually effective in curing a particular disease. In such case laboratory analysis are suggested for further investigation on effective and safe use. Although the folk wisdom of medicinal plant is now gradually disappearing with the rapid indiscriminate devastation of forests. As per the present study, the recovery of diabetes through natural sources is of high importance having no side effects. Diabetes mellitus which is a disorder of the carbohydrate metabolism, glucose cannot enter the cells but remain concentrated in the blood. Thus the cells survive in the midst of plenty. The increased amount of sugar in the blood also brings about adverse changes in the membrane transport system of the cells. The blood circulation is also impaired. A net result of all these changes is that opportunistic bacteria etc. flourish and firstly skin affected easily as it is the most easily accessible part of the body. Then affected the kidneys, eyes, nerves etc. Initially skin affection symptoms occur in sexual organs, toes etc. Increased incidence of skin infection is only one of the group of complications that may set in due to diabetes. All the above mentioned herbal medicines are very effective to subside diabetes. For diabetes inevitable foods are lettuce, coconut, soyabean, camels and goats milk, curd, whey, sufficient water and often starvation.

3.22 COWS (*BOS INDICUS*) MAKE MEDICINE FOR MANKIND

Products of cows has multiple properties and is being therapeutically used since ancient time. Its antibacterial, wound healing properties are promising.

Keywords : Wound healing, cowpathy, panchgavya

Introduction

Cowpathy is a system of medicine for human or veterinary ailments, are treated with products of cow. Panchgavya which is derived from cow i.e. milk, ghee, dadhi, dung and urine. Indian priests used Panchgavya, a sacred liquid mixture served to the people in worship-place for a long time. In recent years, interest has been generated among the folk doctor for the use of cow products because in allopathic treatments particularly there is development of resistance to microorganisms and side effects. So the ancient alternative system of medicine gains momentum in India and broad. Generally bacteria have to be destroyed by phagocytic sytem (the body's own defence mechanism). During the last decade it has been observed that due to enhancement of pollution the total body system is gradually abolished.

Result and Discussion

Recent exploration showed that cow urine enhances the immuno status through activating the monocytes and augmenting their engulfment power. It is too much effective for its use in T.B. patients as it has bioenhancing properties thereby reducing the cost of treatment and its duration. Cow urine alongwith antibiotics also prevents the bacteriostatic conditions in the wounds (Puri 1970). As cow urine has antioxidant properties and thus it neutralizes the oxidative stress produced in body through action on free radicals and thereby retard senescence by Panchgavya. So it has gerontological importance. It is also the usual drug as pregnancy initiator in women. It has been found to repair the damaged DNA and thus it is very effective for cancer therapy. Due to enhance utilisation of pesticides in crops as a consequence apoptosis (cell suicide) causes in lymphocytes and tissues through fragmentation of DNA. Here the preventive curing is also the cow urine. It provides better protection alongwith vaccination in poultry birds, and increases egg production. Regular intake of whey (skimmed milk) a healthy drink can be recommended as it has preventive power like antibiotics against common ailments but no side effect. Skimmed milk is Ca-P-Fe riched but fat free useful drink. To increase the beneficial quality of curd add some amount of whey at the initial stage during curd preparation and used it in diarrhoea. To prepare curd for hyperglycemic (diabetic) patients hung the curd by cloth for 30 minutes. Then add milk in curd according to the requirement. Urine of indigenous Indian cows are very effective. But almost nil or less effective by the urine of crossbred, exotic cows. After addition of neem leaves juice in cow urine the mixture act as a wonderful ecofriendly biopesticides which do not accumulate in the food-chain (i.e. biodegradable) and thereby no risk for biological magnification. The dried dung is burnt and ash is applied externally to treat urticaria, castrated wounds, tooth etc. From the above mentioned useful application it is logical to conclude the reasons behind the Hindus God Culture associated with cow. There is a myth among the tribals that the power order of milk products are as follows : whey > curd > cheese > milk.

Conclusion

According to WHO, the twentieth century's the wonder drugs "antibiotics" will not remain useful and become almost ineffective within 20 years. Then one has to think about the alternative therapeutic approaches. Regardless of scientific validation, tribal people of Bengal are still using and getting benefit of it from the time immemorial. Around 70% of the healthcare needs of India is still being catered by the traditional system of medicine from the natural resources like cowpathy.

3.23 PEST MANAGEMENT PRACTICES THROUGH BIOCIDES

Abstract

Due to increased demand of bio-natural processes the author has tried to unveil some eco-friendly biocides which are cheap, easily available.

Keywords : Pest, eco-friendly, biocides

Introduction

Effects of destruction of ecological balance by indiscriminate utilization of inorgance fertilizers, insecticides, pesticides, fungicides causes hazardous to living beings. Foliar application of leaf extract on the plant's growth was considerably enhanced while disease incidence was markedly reduced by biocides (Chattopadhyay and Maji 1976).

Materials and Methods

The study was carried out repeatedly in the authors field for 3 consecutive years. The household practices and biocides may be the alternative against hazardous effect of insecticides and fungicides. 20 kg freshly collected cow-dung mixed in 200 litre water and the supernatant diluted by 50 litre water which is effective in rice field against pest than the 2 days old cow urine.

Holistic approach or cultural practices

500 gm rasun (*Allium sativum*, Liliaceae) crushed, soaked overnight in 5 litre Kerosine and then filtered. In this supernatant add freshly prepared 100 gm chilli (*Capsicum frutescens* L. Solanaceae) paste alongwith 1 litre soap powder solution. Sprayed this mixture alongwith 40 litre water in rice field. This biocide is very effective against catterpillar, plant hopper, rice hispa, stem borer etc.

Foliar spray of 50 gm seed dust of ata (*Annona reticulata* Annonaceae) in 1 litre water act as insecticide. Even the paste of leaves also act as insecticides on chilli, brinjal, tomato, gram, pea plants. Tender leaves of ata also used orally as anticancer therapy.

Seed dust of kolkey-phul (*Thevetia nerifolia, Apocynaceae*), Shakalu *Pachyrhizus angulatus*, Papilionaceae), root dust of patol (*Trichosanthes dioica*, Cucurbitaceae), leaf extract of bera (*Ipomoea fistulosa*, Convolvulaceae), gulancha (*Tinospora cordifolia*, Menispermaceae), gandal (*Paederia foetida* L. Rubiaceae), nishinda (*Vitex negundo*, Verbenaceae), nayantara (*Vinca rosea*, Apocynaceae), Iswar mul (*Aristolochia indica*, Aristolochiaceae) act as biocide. 500 gm hing (*Ferula asafoetida*, Umbelliferae) dust soaked overnight in 1 litre Kerosene and then filtered. Then in this supernatant 50 gm chilli paste mixed in 1 litre water and made a volume of 100 litre hing-chilli biocide.

10 gm hing boiled in 5 litre water, 100 ml supernatant mixed in 2½ litre water as a preventive mixture against virus.

5 kg neem leaves soaked for 7 days in 20 litre water. From this 5 litre supernatent boiled with 10 gm hing to prepare the biocide.

Ash/wood-coal dust and fine sand is very effective as insecticides. Even the floor of the tribal's hut is sprinkled with wood ash which prevents the entry of unwanted insects.

Conclusion

Prior to any foliaceous spray, foaming detergent powder should be mixed for lusture and the spray will be conducted during morning. For foliar application of biocides, taste and post harvest storage duration period of different rice, brinjal, tomato, ladies finger, pea etc. has been increased without any deleterious effect.

The above mentioned biocides could be economically viable alternatives of conventional synthetic pesticides and be utilized on large scale.

3.24 INDIGENOUS METHOD FOR LUSTRY BUILDING BY NATURAL COMPONENTS

Abstract

Potential value of some easily available natural components used by the age-old masons in South-rardh for lustry buildings, temples, mosques, churches etc. and the process of making the same have been discussed in this paper for further exploration of this age-old indigenous process which is being gradually diminishes.

Keywords : Rardh, natural lime stones, indigenous practice, laterite.

Introduction

Natural lime stone is an excellent component for buildings, temples etc. obtained from Bankura, Purulia and Paschim Medinipur districts have been used since the time immemorial. The mixed components yields a very luxurious, showy medium for ornamenters and designers which can protect from salinity. The natural as well as plant based elements have been effective till now. Modern researchers have also realised the ecofriendly effect of the process.

Materials and Methods

For preparing materials for construction, lime stone is generally collected from hills. Usually for making the base of buildings, temples etc. lime stone, slaked lime [$Ca(OH)_2$] and laterite soil should be mixed in 3:1:2 ratio.

Usually to make the walls, laterite soil and slaked lime mixed to be in 2:1 ratio. In both the above stages egg-yolk, sago (*Metroxylon sagus* Rottb, Arecaceae), fibre pieces of jute (*Corchorus capsularis*, Tiliaceae) and molasses (*Saccharum officinarum*, Poaceae) should be piled in the mixture just prior for its utilisation.

For the preparation of roof two vessels has selected : 1) In the first vessel 1.5 kg Methi (*Trigonella-foenum-graecum* L, Papilionaceae) and 1.5 kg biri (*Phaseolus mungo*, Papilionaceae) soaked for 7 days. In another vessel laterite soil, slaked lime and lime stone mixed in 3:1:2 ratio. The mixture soaked for 7 days. Now 2 vessels contents piled properly. Just prior to utilisation jute pieces, molasses and khaier (*Acacia catechu*, Mimosaceae) mixed in 1 : 2 : 1 ratio along with some amount of yolk, gum of green bel (*Aegle marmelos*, Rutaceae) and sago. Molasses and khaier help to coagulate the mixture instantly.

Conclusion

It is an age old, popular, competent, lustry, designophilic and comfortable during summer months though the process is laggard.

3.25 ECOPHILIC FUELS FOR RAIL ENGINES AND AUTOMOBILES

Bio-diesel the products of plants derived from copaiba latex (*Copaifea* sp) of Brazil the giant diesel Yielding tree which can supply 40 litres diesel/year, and *Jatropha curcas* (amlagota, bag-varenda) seeds yields 1 kg oil/year. Bio-diesel production varies with the qulity of seeds. In India *Jatropha curcas* (Euphorbiaceae) generally found in everywhere. Indian Railways have so many barren lands where these plants can be grown to minimise mineral diesel import. Under South-Eastern Railway the Kharagpur division will be the pioneer and sole survivor if they exert eco-friendly alternative fuel-The Bio-diesel Train. In Kharagpur there are 78 diesel engines which requires daily one Lakh litre (100 Kilo-litre) mineral diesel. If bio-diesel used in these engines the cost will be reduced to half. The refining procedure will be executed by Haldia Refinery. The cost of bio-diesel is Rs. 15/- per litre. Successful Trial venture with the biodiesel was conducted during December, 2002 in Swarna-Jayanti Jana Satabdi Express train between Delhi-Amritsahar. Two years will be the sufficient to get bio-diesel from the seed of planted adult *Jatropha*. Each plants generally yields 3 kg seeds after 2 years. A regular movement of bio-diesel train in the south Eastern Railway will be found within 2006 under the leadership of DRM-KGP, if they develop a project on bio-diesel from *Jatropha*.

Conclusion

A dry seed of *Jatropha curcas* Linn yield about 55% of vegetable oil. There are 2 methods of extraction :-(1) Solvent extraction with n-hexane, (2) Mechanical extraction method. Two blends can be prepared for diesel engines by mixing mineral diesel and bio-diesel in the following proportion 1:1; 4:1. Vegetable oil can not be easily used for its high viscosity. It can be used after slight modification of diesel engines. Two years project setting up orchards of *Jatropha* and Conversion of *Jatropha* oil into bio-diesel optimise the income from such plantation. The bio-diesel has less CO_2 emission, low sulphur, high O_2 contents and lubricating properties. High yielding varieties of *Jatropha* from Mexico will now be cultivated in Indian Railway barren lands.

Jatropha curcas cures for yellow fever, piles and laxative. A good source of potash for soap making and may be used as bio-fertilizer. On behalf of Kisan Herbal Garden in Desra village of Bankura district (P.S.-Kotalpur), W.B. *Jatropha crcas* has already cultivated as a pioneer effort.

3.26 WOOD-ASH – THE NATURAL PRESERVER

Abstract

The paper focuses the IPTK (Indigenious people's technical knowledge) on natural preserver-The wood ash which is traditionally used in health care practices and in agriculture which have antiseptic and curative properties.

Keywords : IPTK, Natural Preserver, Wood-ash.

Introduction

IPTK on natural product are immemorial old activities carried out by the lay persons. Looking at Indian history peoples have engaged themselves in a constant struggle with disease and pests are now gradually diminishing.

Observation

The author observed various beneficial achievements by the wood-ash such as :

1. Traditional optimise tooth powders were prepared by utilizing the wood ash alongwith edible salt for clean and

disease free tooth. Besides ash of groundnut shell mixed with neem and haritaki (*Terminalia chebula*) wood is very effective.

2. Freshly collected disease free and healthy tomatoes which are ripe but not over-ripe and soft can be preserved in wood-ash upto 3 months.
3. In castrated domestic animals like goat, ox, dog etc. wood-ash is used in the castrated wound.
4. The ducks which are suffering in arthritis must be kept in a big wood-ash container for 3-7 days for its recovery.
5. Typhoid and pneumatic calves and cows become disease free if laid down over wood-ash in cow-shed.
6. All pulses and cereals can be free from pests by mixing wood ash with the grain. Besides vegetable crops especially gourd, ridge-gourd, cucumber can be escape from beetle and insect pests by dusting with ashes.
7. All utensils become very clean after rubbing with wood ash.

Conclusion

Common people at large are still adopting some of the common indigenous technical knowledge developed by the ancestors based on their long experience through easily available household resources. There is a growing demand to develop global database on systematically evolved indigenous people's technical knowledge (IPTK) which have no side effect. Although allmost IPTK techniques are perfect, rather it intends to highlights these eco-friendly techniques.

3.27 NATURAL WAY OF SUGARCANE CULTIVATION THROUGH DEW-TRAPPING PROCESS

Abstract

Potential source of Dew-drops are used by the farmers of Purulia District in the droughty areas to cultivate sugarcane (*Saccharum officinarum*, Poaceae). Dew-trapping is an excellent cultivation process.

Keywords : Dewy soil, anthesis, tribal, heterozygous

Introduction

Though most of the graminaceous plants underwent reproductive propagation yet in sugarcane vegetative propagation has been found. Dew-drops and fog are formed by natural process during winter months from October to February. Purulia is the tribal inhabitant district of West Bengal and lying between 22° 46′-22° 38′ N and 86° 36′-87° 46′ E parallel. Here the average annual rainfall is about 1200 mm and a temperature ranging about 25°C-50°C. In spite of scarce rainfall and deficient in irrigation facility profuse sugarcane cultivation has been adopted by the farmer's of Sircabad through Dew-trapping (i.e. randa process) since the time immemorial, probably it is in use since 100 years ago introcued by late Nidhiram Dutta of Sircabad under Arsha block which is now an indigenous popular process for developing sugarcane. The present work aims at separating the role of randa process from the traditional cultivation.

Materials and Methods

The land is neatly and thoroughly prepared by 2 deep ploughings followed by clod-crushing and leveling by ladder. Planting is generally done during March i.e. with the advent of warm weather. As the cultivated varieties of sugarcane are heterozygous and do not breed true. The plants are therefore propagated by vegetative means. In-plant germination of sugarcane healthy seeds or sets propagated (one can induce bud germination on the sugarcane plant itself, cut the sets i.e. nodes and sow the early germinated nodes) usually used for cultivation. The field is ploughed deeply and seeds are placed in the furrows or trenches in lines and are covered with moistened soil previously mixed with rotted farm-yard manure.

Dew-trapping (randa) Process

For loosening the soil ploughing will be done at afternoon and then followed by dew covering by ladder at dawn to get dewy soil. Randa will be executed once in a week from October to February (i.e. 20-25 times in a crop cycle). As a consequence softly-lucid soil has been maintained which are the essential requirements for sugarcane cultivation. Harvesting is done usually from December to January prior to anthesis because profuse germination is sure and uniform planting is done thus saving on labour for replanting. 2000-3000 acre land has been cultivated by the process and a net profit has been achieved rupees 10000-15000 per acre.

Conclusion

Though today sugarcane is widely cultivated in almost all the moist tropical and semi-tropical regions yet the dew-trapping process of cultivation is an indigenous and innovatory one of Sircabad. In the next year sugarcane is grown in rotation with paddy, pulses to maintain the soil fertility.

3.28 TAILORING OF HAULMS SERVE AS A NATURAL REMEDY AGAINST "LATE BLIGHT OF POTATO" AND NATURAL RESOURCE FOR BIO-FERTILIZER

Abstract

During the post tuberization period (75 days onwards) of potato plants the haulm cuttings can be executed at 90-95 days having no side effect on tuber yield. Besides this method is very effective to get rid of from late blight of potato disease in storage condition. Haulm cutting also serve as a bio-fertilizer and minimize the expense of potato cultivation in the coming year. So, potato plants may provide some useful clues, which may help elucidating the mechanism of natural cultivation without utilization of chemical manure in other crops. Evidently, the presence of haulm after 95 days is more deleterious so far for development of disease than its absence.

Keywords : Haulm, tuber, bio-fertilizer, anthesis.

Introduction

Though most of the Solanaceous plants underwent reproductive propagation and are monocarpic yet not all fruit bearing solanaceous plants develop seedling from seed. Interestingly, most of the potato plants under normal condition show vegetative propagation and die vegetatively after completion of tuberization. The present work aims at separating the role of haulms at post tuberization period.

Materials and Methods

Sliced tubers of *Solanum tuberosum,* having well defined vegetative buds weighing 50 gm were surface sterilized with 0.1% (w/v) $HgCl_2$ for 1 min and then washed thoroughly in tap water. Seed tubers were then sown during the month of November in the field in line (10 cm deep) on the ridges (at a spacing of 45 cm between rows and 15 cm between seed tubers) previously moistened and prepared with farm-yard manure. These were then covered with soil. Sprinkling of water was done on the seed bed at regular intervals until sprouting. Two weeks after emergence of the plant, the soil between the lines was transferred around the plants in such a way that the plants were ultimately raised on straight ridges. Appropriate watering was done between the ridges at an interval of 10 days. Tuberization of the plants started at the plant age of 35 days. Anthesis occurred at the plant age of 45 days. Among the plants having uniform growth rate in order to observe development of disease (late blight) symptoms under storage condition physical manipulations (excision of haulms i.e. green aerial shoots) were done at the age of 95 days on 50% plants. Irrigation will be stopped before 10 days to removal of haulms.

Results and Discussion

Physical manipulations were done at 95 days when there is no positive role of haulm exist on tubers yield and its maturation. Tailoring of haulm at 95 days indicate that translocation towards tubers (Ghosh and Biswas, 1989, 1991) is not required or hampered possibly due to maturation of the tubers. Excised plants having same tuber weight, the tubers, which were stored separately, found disease free even in post storage period compared to normal intact plants. At 95 days the proper maintenance of tuber yield due to a balanced nutrient exhaustion (Moorby and Milthrope, 1975, Biswas and Ghosh, 1999) out of the storage organ. Thereby excision of haulms produce disease free tubers. Due to lack of haulms no fungicide spray is required and excised haulms at the same time may used as bio-fertilizer in the next season. The results also emphasize that at post tuberization stage (after 95 days) haulms served to tubers as susceptible to disease development (i.e. caused deleterious effect to tuber) though source-sink variation (Biswas and Ghosh, 1999) during whole plant senescence which succumb to death in vegetative as well as in reproductive condition.

3.29 THE NATURALLY CULTIVATED AND PROCESSED GROUNDNUT MILK IS THE ALTERNATIVE OF COW MILK

Abstract

Potential value of easily available natural vegetative groundnut milk used by the Villagers of South Bengal and process of making milk have been discussed in this paper for further exploration.

Keywords : Natural vegetative milk, South Bengal, Brick Powder.

Materials and Methods

Though the ground nut originated in the North : West of Brazil, now cultivated widely. In *Arachis hypogaea*, 2 types of flowers occured : (1) Cleistogamous flowers situated below the leaves mature into fruits, while, (2) Aerial flowers (Chasmogamous) are abort and do not mature into fruit. The ground nut is a palatable valuable food and oil crop. Its seed content large amount of fat (55-60%) and protein (40-45%). Seeds are used in manufacturing vanaspati, cheese, butter, milk, condensed-milk, posset etc. I used before planting hard bricks powder and clay powder in 1:2 ratio to reduce toil as well as to ensure maximum out put (pods), free from kohiwara disease and the roots do not cut during plucking. It ensures that the plants grow well in lateritic soil. Surface sterilized and water soaked seeds were sown in the first week of April in lines at a depth of 2 cm in the moist lateritic soil (mixed with brick powder and clay). Two weeks after emergence plants were thinned to 30×30 cm between plant and rows. Loosening of the soil and covering of the newly formed begs were done at periodic intervals. Watering was done as and when required. Fruits were setting from the plant age of 59 days and matured at the age of 130 days. Harvesting was conducted on 137 days. To control rat, goat's hair hidden in sweet-ladoos placed in the field for choking their respiratory tract and intestinal organs.

For preparing ground nut milk 1 kg seeds were soaked for overnight in cold water and then dehusked. Seeds are pounded or grinded in mixie machine and followed by mixing and stirring among 10 litre hot water to yield the natural milk. Always the proportion of ground nut powder and water will be 1:10. To meet the demand of cow milk, ground nut milk is the best for all age groups of human.

Comparison between food value of 1 litre cow milk and ground nut milk

Cow milk	*Ground nut milk*
670 Cal.	549 Cal.
Protein 32%	40-45%
Fat 41%	35-60%
Carbohydrate 44%	20-22%
Cost Rs. 13/-	Rs. 4/-

Ground nut milk is an excellent natural milk cheaply available from underground or geocarpic fruit. This milk preparation can be processed commercially through scientific high technoloy. As a pioneer report milk has been extracted in the village Panchisbetia under North Kathi assembly with the co-operation of Balagaria Central Co-operative Bank. Maximum ground nut cultivation is in Egra, Kanthi, Panskura, Patashpur, Block. Ground nut yield i.e. milk production can be increased through defloration of aerial flowers as they have some deleterious effect on the yield (Ghosh and Biswas-1989) of ground nut and they are the supplier of senescence signal. At the same time cultivation cost can be minimised through tailoring of aerial flowers.

3.30 SERIPHILIC PLANTS IN AGRO-FORESTRY

Keywords : Seriphilic plants, agro-forestry, biocidal.

Mulberry, Tasar, Eri and Muga with their numerous species and verities have enriched silk production. The host plants of mulberry, tasar, eri and muga are though present in large numbers still they are hidden. The present paper aims to unveil them alongwith some uses. Traditionally, kantha stitches on silk were done on old sarees, dhooties, bags, kurta, punjabees are very lucrative in Bengal.

Silk the product of silkworm derived from cocoons have enormous uses. Silk production varies with the quality and quantity of leaves they fed during the larval stage. For the management of silkworm and getting silk throughout the year silkworm keepers can select multivoltine according to the availability of different seriphilic plants. Some plants suitable for sericulture have been enumerated below for the farmers. For mulberry propagation and cultivation suitable varieties are : *Morus indica* L, *Morus serrata, Morus levigata, Morus alba.* These plants are suitable for *Bombyx* species.

For *Antheraea mylitta* (Tasar) : *Terminalia arjuna* W & A, Arjun (Combretaceae), *Terminalia tomentosa* W & A, Asan (Combretaceae), *Ficus carica* L., Dumur (Moraceae), *Shorea robusta* Gaertn, Sal (Dipterocarpaceae), *Butea monosperma* Taub palash (Papilionaceae), *Ziziphus jujuba* Lamk, Kul (Rhamnaceae), *Cajanus cajan* L., Arhar (Fabaceae).

For *Antheraea assamensis* (Muga) : *Machilus bombycina,* Some (Lauraceae), *Litsea polyantha,* Sowalu (Lauraceae).

For *Philosomia ricini* (Eri) : *Ricinus communis* L., Rehri (Euphorbiaceae), *Carica papaya* L., Peypey (Caricaceae), *Jatropha hastata,* Bharenda (Euphorbiaceae), *Plumeria acutifolia* Poir, Golancha (Apocynaceae).

Feeding of tender leaves to silkworm increased cocoons qualities. Quality of mulberry and other seriphilic plants leaves influence the economic characters of cocoons. Through the feeding experiment on multivoltine silkworm hyrids, it was observed that unlike conventional methods, feeding of tender leaves particularly during V stage resulted in higher cocoons weight. Beisdes scarce feeding (2 to 3 times) to silkworm is beneficial so far as economic efficiency is concerned. Savings labour amounts to 25-50% compared to 4 times feeding. Application of foliar spray of water hyacinth (*Eichhornia crassipes* solms) leaves extracts and leaf leachate of mango (*Mangifera indica*) on seriphilic plants caused luxuriant growth of leaves. Besides application of several biocides to seriphilic plants has been found useful to control hazardous effect of insecticides and fungicides thereby enhanced plant growth and reduced disease incidence. Such biocidal plants are : *Allium sativum* L., Rasun (Liliaceae), *Vinca rosea* L., Nayantara (Apocynaceae), *Datura stramonium* L., Dhutura (Solanaceae), *Zingiber officinale* Rosc, Ada (Zingiberaceae), *Curcuma longa* L., Halud (Zingiberaceae), *Azadirachta indica* A. Juss, Neem (Meliaceae).

3.31 THE NEW SWEETY NATURAL SAFE BEVERAGE YIELDING PLANT OF BENGAL

Introduction

The body is the product of food and drink. Diseases occur as the result of abnormal diet. Modern peoples are depended on fast foods and drinks. So different types of herbal health drinks can fulfill nutritive and preventive requirements of human body. *Piper betle* L (betel-vine) of piperaceae is endemic and grown abudantly in Bengal. Though it is a very popular chewing material, its dark green leaves have hot astringent taste. The plant is grown under shaded conditions to cause an increases in the size of the leaves and the longivity of plants.

Keywords : Beverage, fast food, depetiolated, dechlorophyllated

Materials and Methods

Freshly collected healthy depetiolated leaves washed repeatedly at first in cold water and then in 60-80°C hot water to remove germs. Extracted the juice by crusher machine from the leaves and then dechlorophyllated by the centrifused mechine. Add in this mixture coarsely crushed rest all of the ingredients in proportionate ratio to make the drink healthy and tasty which have no adverse side effect. The ingredients are citric acid, Jeera/cumin (*Cuminum cyminum* L, Umbelliferae) ginger (*Zingiber officinale* Rosc, Zingiberaceae) Pepper (*Piper longum* L., Piperaceae), Rock salt, ascorbic acid and sugar.

These ingredients made the healthy beverage or Pan-drink. When through high pressure Co_2 is passed and mixed in the beverage then it will be carbonated and named as 'Soft-drink'. Ten thousand leaves of pan can yield more less 2-2½ thousand litre drink whose cost will be around 6-7 thousand rupees in the market.

Conclusion

Generally people of Bengal procure and habitually carry the betel-vine leaves with them and chew the same whenever the digestion may seek relief. Besides treating indigestion the leaves have been used locally to stimulate the appetite and relieve flatulence and to act as a general tonic. As there is no doubt about its palatability, so it will be the most wanted potential herbal health drink. Though *Piper betle* is a traditional medicine for digestion yet its newly herbal drink is good for calory value, blood purifier, nerve-stimulation, reduction of body fat, cold, cough, dysentery, fever, diptheria, toothache, gas troubles, bad smell of month and even as an antidote against different germs (Vedavathy *et al.* 1997, Sethi 2004, Vedavarthy 2004). In near future pan-drink can be used in ice-cream, chocolate and as a flavouring agent. This is poor mans' drink for Bengal, and is served especially during special occasions to the people. Due to full of antioxidant it can delay senescence.

Acknowledgement

The author is grateful to Kallol Nanda and Sandeep Mitra of Midnapore for their active effort to establish safe-beverage industry at Haldia in near future.

3.32 HITHERTO UNTAPPED PLANTLORE FROM BENGAL

Abstract

Plants are important natural source of medicine since the time immemorial, still some are remain unscreened for their bioactivity in various ailments. The present finding is aimed to discuss biological activities of plants in different living beings.

Keywords : Bioactivity, Biocide, Vernacular.

Introduction

Since time immemorial West Bengal has been regarded as a resource reservoir of ethnobiology. However, still large number of plants and animals used by local inhabitants of this region are little known. The folk utilization of several plants is on the verge of decline. In view of this exploitation of well known plants, animals, search for natural resources and conservation of folk herital knowledge, an attempt has been made to study, the ethnobiological aspects from this region. Since plants species is playing an important role to cure ailments (Brown 1939, Krei 1964, Lezy 1964, Chettri et al 1992, Mondal and Chauhan 2000, Ghosh 1999, 2002, 2003).

The Study Area

Medinipur district is situated in the south-western corner of the state between 21° 36′ 40″ and 20° 56′ 40″ N & 86° 35′ 22″ and 88° 31′ 30 ″E longitude. Its areas is 13,000 Km^2. Forest cover about 1700 Km^2. The Lodhas (No. about 12000) inhabit in the northwestern part of the district. Santals are about 38,000 distributed throughout the district. Here the Santals and Lodhas are the main tribals. Besides the other tribals are Bhumij, Koras, Mahali, Mech, Munda etc.

Methods

Ethnobiological surveys of the tribes and folklores were undertaken during the period of 1995-2005.

During the field trips personal interviews were taken with chieftains, old and experienced peoples and folk doctors for documenting ethnobiological medicines. For documenting the medicinal properties the herbalists of every anchal were accompanied with me in the field and forest to identify the plants used by them in different ailments. The informers were interviewed at the spot, using a questionaire. As a strategy to confirm the information it was rechecked by me and also with other herbalists. Repeated queries were also made to get the data verified and confirmed. Vaucher specimens alongwith taxonomic determination made by the competent authority and were deposited in the herbarium of botany department (Ramananda College, Bishnupur).

Results and Discussion

The tribals have certain beliefs about plants :-

1. Plants should be collected for Medicinal uses in the morning.
2. Roots are considered more effective compared to aerial parts.
3. Fresh plants before anthesis are more effective than dried plants.
4. Plants with latex have medicinal value.
5. Bark should be collected from the east side of the plants.
6. Herbal medicine is more effective if peppar is added.

Violation of the above mentioned beliefs or taboos can make the medicine ineffective. We neither affirm nor deny the efficacy of any medicinal plants included in this paper. How the ethnobiological species is playing a prominent role to cure has been described in the table 3.32.

3.33 SAPLING OF BAMBOOS THROUGH NODAL CUTTINGS IN BENGAL

Abstract

To achieve self-employment commercial cultivation of bamboos through nodal cuttings is easier and have great demand for its edible shoots in Bengal.

Keywords : Nodal cutting, edible shoots, monocarpic, culm.

Introduction

It has a peculiarity of flowering and seeding only after a long vegetative phase. In the traditional way of cultivation it is difficult to produce bamboos sapling. Though it is a monocarpic plant yet it is not grown from seeds. Most of the graminaceous monocarpic plants grown through reprodctive propagation. Its wide distribution takes place through vegetative propagation. Its rapid rate of growth and easy handling make bamboo an ideal material for countless uses among poor tribals. Still bamboo shoots are an important daily food for them. No attempt has yet been made so far to evaluate and elucidate the bright possibility of nodal cutting among farmers.

Materials and Methods

Keeping in view among the so many genera and species, only a few are grown commercially for their shoot, salinity stress, preventive power against pest, lustycity, adaptability to different climates. The occurrence as well as cultivation through nodal cuttings in Bengal's diverse environmental conditions alongwith their cultivation prospects have been discussed in this paper.

Viable, surface-sterilized and soaked (in root initiating hormone) nodal cuttings from the healthy mother plants allowed to roots on sandy bed, previously prepared with rotted organic manure at shady place during the rainy season or at the advent of full

Table 3.32

Ailments	*Vernacular Name*	*Botanical Name*	*Family*	*Parts Used*	*Uses*
Constipation, Stomach pain	Papaya	*Carica papaya* L.	Caricaceae	Fruit	Fed an unripe fruit daily in constipation whereas ripe fruit is taken in stomach pain.
Dimness of Vission	Tela Kucha	*Cephalandra indica* Naud	Cucurbitaceae	Leaf	Leaves are fried in ghee/oil and taken daily orally to improve eye sight.
Constipation	Rangalu	*Ipomoea batatus* L.	Convolvulaceae	Root	Tuberous root fed daily claimed as laxative.
Night blindness	Noniasag	*Portulaca oleracea* L.	Portulacaceae	Aerial Part	This vegetable is nice against night blindness.
Fish catching	Hazarmani	*Phyllanthus urinaria*	Euphorbiaceae	Leaf	Decoction used in Jaundice and fish poison.
Dysentery	Peyara	*Psidium guajava* L.	Myrtaceae	Leaf	Juice from the crushed leaves taken orally to cure dysentery.
Night blindness	Bhang	*Cannabis sativa* L.	Cannabinaceae	Leaf	Through smoking or ingested a crude tincture were apparently able to see in darkness.
Leukaemia	Piaj	*Allium cepa* L.	Liliaceae	Bulb Powder	Smear the paste

contd...

Table 3.32 – *contd...*

Ailments	*Vernacular Name*	*Botanical Name*	*Family*	*Parts Used*	*Uses*
Diarrhoea	Bel	*Aegle marmelos* L.	Rutaceae	Fruit Pulp	Add a pinch of black salt in this mixture (5:1) and fed to animal and human.
	Khayer	*Acacia catechu* L.	Mimosaceae	Resinous extract	
Toothache	Narikel	*Cocos nucifera* L	Arecaceae	Shell	Ash of shell alongwith powder of curry leaves use regularly with edible salt.
	Karipatta	*Murraya koenigii* Spreng	Rutaceae	Leaves	
Diarrhoea	Karipatta	*Murraya koenigii* Spreng	Rutaceae	Leaves	Fed leaf extract to the affected animals and the patients.
Diabetes	Karipatta	*Murraya koenigii* Spreng	Rutaceae	Leaves	Fed leaf extract to the patients.
Seedling rot of Tomato	Baganbilas	*Bougainvillea glabra*	Nyctaginaceae	Leaf	Prior plantation seeds will be soaked for 6 hrs in the juice of leaves.
Recurrent breeding preventer as a cooling agent	Isabgol	*Plantago ovata*	Plantaginaceae	Seeds	Fed to the cattle for 3 days alongwith castor oil in 1.5 ratio.
Toothache	Isabgol	*Plantago ovata* Forsk	-do-	Root	Roots are used.

contd...

Table 3.32 – *contd...*

Ailments	*Vernacular Name*	*Botanical Name*	*Family*	*Parts Used*	*Uses*
Biocides for lice, ticks, fungas	Methi	*Trigonella foenum-graecum* L.	Leguminosae	Seeds	Seeds are soaked in water overnight to make paste, smear the paste on the body of buffaloes and allowed them to remain immersed in pond. Then fishes attracted and acted as a biological controller.
Rheumatism	Grey pumpkin (Chalkumra)	*Benincasa hispida* (Thunb.)	Cucurbitaceae	Fruit	Feeding for 7-10 days once daily to bullock.
Spermatorrhoea, conception	Aswatha	*Ficus religiosa* L.	Moraceae	Recep-tacle	Fed it to stop conception.
Retention of placenta	Shimul	*Bombax ceiba* L.	Bombacaceae	Twigs	If placenta is not expelled within 6-8 hrs after parturition fed 1 kg decoction to animals.
Sexual debility	Shimul	*Bombax ceiba* L.	Bombacaceae	Root	5gm roots powder fed daily.
Joint sprain and pain	Nishinda	*Vitex negundo* L.	Verbenaceae	Leaf	Ground the leaves alongwith lime and ant-hill to make a paste for external use in cattle and man.

contd...

Table 3.32 – *contd...*

Ailments	*Vernacular Name*	*Botanical Name*	*Family*	*Parts Used*	*Uses*
Constipation	Lanka	*Capsicum frutescens* L.	Solanaceae	Fruit	Fed mix powdered of 5-6 chillies in a litre of water thrice daily to bovines.
Diabetes	Safeda	*Archras sapota* L.	Sapotaceae	Root	Drunk 1 teaspoonful juice daily.
Liver trouble	Kali cha	*Camellia sinensis* L.	Theaceae	Leaf	Drunk a glass of tea to sick animals.
Constipation	Chola	*Cicer arietinum* L.	Fabaceae	Cotyledon	Fed to animal 250 gm powder alongwith edible salt.
Skin disease	Narikel	*Cocos nucifera* L.	Arecaceae	Fruit	Ash of pericarp rubbed externally both in human and bovines.
Skin disease	Dhutra	*Datura metel* L.	Solanaceae	Seed	Paste made by powder of seeds mixed with copper sulphate and coconut-oil smear externally on the skin of camel, horse, sheep, goat, cows etc.
Bone-fracture	Bot	*Ficus benghalensis* L.	Moraceae	Bark	Paste of bark used as balm for cattle.
Bone fracture	Mehandi	*Lawsonia inermis* Lalle	Lythraceae	Leaf	Paste of leaves used in cattle, camels etc.

contd...

Table 3.32 – *contd...*

Ailments	*Vernacular Name*	*Botanical Name*	*Family*	*Parts Used*	*Uses*
Lactation	Sajina	*Moringa oleifera* L.	Moringaceae	Leaf	Fed the animals alongwith molasses.
Fever	Rehri	*Ricinus communis* L.	Euphorbiaceae	seed	Caster oil is given to goat, sheep
Shoulder pain	Kul	*Zizyphus mauritiana* L.	Rhamnaceae	Root	Smear the paste on shoulder regularly as balm.
Mouth disease	Sajaru (Porcupine)	*Hystrix indica*		Squills	Fumigated for curing mouth diseases of cattle.
Jaundice	Shialkata	*Argemone mexicana* L.	Papaveraceae	Leaf	3 drops of juice mixed with a cup of cows milk is taken once daily.
Dandruff	Piaj	*Allium cepa* L.	Liliaceae	Bulbs	Juice of, squeezed bulbs is applied on scalp
Ageing	Nalakashina	*Smithia germiniflora* Roth	Fabaceae	Whole plant	Decoction is useul as ageing retardant and stomach ulcer.
Sexual potency	Gandharaj	*Gardenia aqualis* Guale	Rubiaceae		Juice of leaves improve sexual potency.
Contraceptive	Kanchphal	*Abrus precatorius* L.	Fabaceae	Seeds	Pills of 1 mg with jaggery fed 1 daily from 5th day of mens consecutively for 5 days.

contd...

Table 3.32 – *contd...*

Ailments	*Vernacular Name*	*Botanical Name*	*Family*	*Parts Used*	*Uses*
High blood pressure	Dhaniya	*Coriandrum sativum* L.	Apiaceae	Leaf	Juice is used to bring down high blood pressure.
Snakebite	Apang	*Achyranthes aspera* L.	Amaranthaceae	Root	Root is crushed in water and paste applied in the wound.
Hydrophobia	Apang	*Achyranthes aspera* L.	Amaranthaceae	Seeds	Seeds are used as tomic in hydrophobia
Snake-bite	Sarpagandha	*Rauvolfia serpentina* L.	Apocynaceae	Root	Root is crushed in water and given orally as well as applied on the site of snake bite.
Snake-bite	Bel	*Aegle marmelos* Corr	Rutaceae	Root	Paste of root applied on the wound.
Snake-bite	Sona	*Oroxylum indicum* vent	Bignoniaceae	Fruit	Paste used externally.
Snake-bite	Gulancha	*Tinospora cordifolia* L.	Menispermaceae	Entire plant	Infusion will be taken.
Snake-bite	Somelata	*Sarcostemma acidum* voigt	Asclepiadaceae	Root	Infusion taken in snakebite and dog-bite.
Warding off snakes, reptiles and rodents	Halud	*Curcuma longa* L.	Zingiberaceae	Turmeric powder	Fumigate the rooms with turmeric fumes which also act as sterilizer in epidemic of pox, cholera, T.B.

contd...

Table 3.32 – *contd...*

Ailments	*Vernacular Name*	*Botanical Name*	*Family*	*Parts Used*	*Uses*
Snake-bite	Lajjabati	*Mimosa pudica* L.	Mimosaceae	Root	Root extracts against cobra bite
Snake-bite	Chaya	*Aerva lanata* L.	Amaranthaceae	Root	Root extract to be used.
Insect bite	Kendu	*Diospyros exsculpta* Buch.	Ebenaceae	Leaf	Smear the leaves on the affected area.
Snake bite	Isharmul	*Aristolochia indica* L.	Aristolochiaceae	Leaf	Juice of leaves is given at interval of half an hour.
		Leucas cephalotes (Roth)	Lamiaceae	Leaf	Paste of equal quantities of the leaves of neem and *Calotropis procera* are used externally.
Diarrhoea	Dhuna of Sal	*Shorea robusta* Gaertn	Dipterocarpaceae	Resin	Fed the powder with lukewarm rice.
	Mouri	*Foeniculum vulgare* Mill.	Apiaceae	Seed	Fed to the cows with molasses.

monsoon. Sprinkling of water was done on the cuttings as and when required. Usual procedure of plant management was followed. For the present study monocarpic bamboos were earmarked in the forest adjacent to Sonamukhi College, Bankura. 2-3 nodes containing culm buds, can be used by placing them upright-horizontally keeping an angle. The sprouted cuttings can be planted in the desired field.

Result and discussion

Following species of bamboos have bright possibility through nodal cuttings so far been achieved.

Flowers gregariously. Thorns and large yellow culms sheath. Its seed rice promote fertility.

Bambusa arundinacea Willd, *B. vulgaris* Schred,

B. tulda Rox, *B. bambos* L., *B nutans* Wall, *B. Polymorpha* Munro,

B. bacoifera Roxb,

Since bamboo seeds are of rare occurrence and have their short period of viability so propagated vegetatively is the best method (Shanmughavel 2004). Selected vernacular name of bamboos are Bombai, Java, Kaldomoni, Barial, Gurivalki, Kanta, Taral, Champak. Among them Kaldomoni, Barial, Valki are fit against salinity environment and stress. As the Java are less-sweety so it is pest resistant and lusty. Through nodal cuttings at a time numerous sapling can be evolved and may be distributed among the silvi-farmers. Conventional propagation is beset with problems of poor seed viability, scanty delayed rooting of traditional vegetative cuttings. Therefore, there is a need for alternative propagation method.

3.34 SENESCENCE RETARDANT AND ANTIOXIDANT POTENCIAL PLANTS

Abstract

National dietary antioxidants like vitamin C, E, Beta (β) carotene delayed senescence. With the advent of ageing some deleterious products accumulates within the organism accelerate senescence.

Keywords : Senescence, ageing, antioxidant.

Introduction

Senescence is a phase of ageing which is degenerative and ultimately leads to death. Though gerontologists want to retard the senescence process through life-saving foods and drinks. Generally nutritionally adequate somewhat deficient in calories full of antioxidants food and drink intake delayed the senescence process (high level of antioxidants retard senescence). Some foods and drinks have free radical scavenging properties (Ghosh 2005). Large number of food and drink with antioxidant activity are being unveiled now.

Materials and Method

Antioxidant activities of different food and drink calculated by using the ferric reducing antioxidant power assay (FRAP assay). Besides, Antioxidant potency were also determined by estimating superoxide dismutase activity (SOD) and catalase activity.

Result and Discussion

High level of antioxidant delayed senescence because it attacks free radicals, the poisonous byproducts of biochemical reactions such as breathing, nutrition, circulation, excretion and other physiological activities. Free radicals are the prime cell killers in connection with or pertaining to ageing, senescence, disease and longevity. The dietary antioxidants stop or neutralize the activity of oxygen free radicals (which is formed during ATP formation as byproduct) thereby increase the longevity and to prevent old age degenerative changes. Antioxidants (butylated hydroxyanisolein) are used to minimize the oxidative destruction of vitamins and essential fatty acids.

List of Plants possessed relatively high antioxidant activity and might be rich sources of natural antioxidants

Vernacular Name	*Parts used*	*Botanical Name with family*
Date palm	Fruit	*Phoenix dactylifera* L. Palmae
Guava	Fruit	*Psidium guaiava* L. Myrtaceae
Pomegranate (Dalim)	Fruit	*Punica granatum* L. Punicaceae
Grape	Fruit	*Vitis vinifera* L. Vitaceae

contd...

Vernacular Name	*Parts used*	*Botanical Name with family*
Haritaki	Fruit	*Terminalia chebula* Retz Combretaceae
Blue-green algae	Fruit	*Spirulina platensis* Cyanophyceae
Amlaki	Fruit	*Emblica officinalis* Gaertn Euphorbiaceae
Orange	Fruit	*Citrus reticulata* Rutaceae
Methi	Fruit	*Trigonella foenum-graecum* Papilionaceae
Tulsi	Fruit	*Ocimum sanctum* Labiatae
Pudina	Leaf	*Mentha piperita* L. Labiatae
Neem	Leaf	*Azadirachta indica* A Juss Meliaceae
Pan	Leaf	*Piper betle* L. Piperaceae
Halud	Stem	*Curcuma longa* L. Zingiberaceae
Anantamul	Leaf	*Hemidesmus indicus* R.Br. Asclepiadaceae
Jasthimadhu	Fruit	*Glycorrhiza glabra* L. Papilionaceae
Bhui-amla	Fruit	*Phyllanthus niruri* L. Euphorbiaceae
Sajina	Fruit	*Moringa oleifera* Moringaceae
Cha	Apical bud and 2 tender leaves	*Camellia sinensis* L. Theaceae
Tomato	Fruit	*Lycopersicon esculentum* Solanaceae
Ada	Modified stem	*Zingiber officinale* Rosc Zingiberaceae.

3.35 NATURAL DYE-YIELDING PLANTS FROM WEST BENGAL

Abstract

The demand of natural dyes gradually increasing for their utility in pots, inks, paints, polymers, textiles beverages among the conscious people as they have no adverse side effect.

Keywords : Natural dye, mordant, synthetic dye.

Introduction

Though Chinese have recorded the use of dye even before 2600 BC but at present there has been increasing interest in natural dyes, as consumers have become well aware of health and ecological hazards related to the use of synthetic dyes (Dassanayake 1978, Ghosh 2003). Common natural dyes evolved from different types of clays, seeds, flowers, leaves, stems, barks, roots etc. Natural dyes vary according to seasonal climante, soil conditions etc.

Materials and Methods

Generally the colour extraction is done usually by powdering the material followed by boiling in water for 10-20 minutes. The yarn of fabric to be dyed is first washed well then heated in the extract at variable temperatures normally for about 30-45 minutes. The creation of a bond between the colouring matter and fibre is called mordanting (a pre-dyeing process to make the fibre receptive to dye). Generally natural dyes require mordant (metalic salts of Al, Fe, Cu, etc.) for ensuring the reasonable fastness of the colour to sunlight. Earthen pots are also suitable to procure basic original colour.

Conclusion

Documentation of traditional knowledge about natural dyes used by weavers, potters, pharmacist, sweet-makers and confectionaries is essential for future generation alongwith establishment of long-term sustainability and dye-yielding substances by commercial cultivation.

Table 3.35

Botanical name & family	*Vernacular name*	*Parts used*	*Uses*
Butea superba Roxb (Fabaceae)	Lata Palash	Root	Dyeing
Morinda citrifolia L. (Rubiaceae)	Indian Mulberry	Root bark	Dyeing (dull red)
Prunus persica Benth (Rosaceae)	Peach	Root bark	Dyeing
Rubia cordifolia L. (Rubiaceae)	Madar	Root, stem	Dyeing

contd...

Table 3.35 – *contd...*

Botanical name & family	*Vernacular name*	*Parts used*	*Uses*
Acacia nilotica L. (Mimosaceae)	Babla	Bark	Dyeing
Bauhinia purpureae L. (Caesalpiniaceae)	Devakanchan	Bark	Dyeing (Purple)
Cassia fistula L. (Caesalpiniaceae)	Badarlathi	Bark, Sapwood	Dyeing (Red)
Casuarina equisetifolia Forst (Casuarinaceae)	Jhau	Bark,	Dyeing (light Reddish)
Ceriops tagal perr (Rhizophoraceae)	Mat goran	Bark	Dyeing (Black/ purple)
Madhuca indica J.E. Gmel (Sapotaceae)	Mahua	Bark	Dyeing (Reddish-Yellow)
Pterocarpus marsupium Roxb (Fabaceae)	Pitsal	Bark	Dyeing (Brownish-red)
Terminalia ariuna Roxb (Combretaceae)	Arjun	Bark	Dyeing (light brown)
Nymphaea alba L. (Nymphaeaceae)	Waterlily	Rhizome	Dyeing Blue
Bougainvillea glabra Choisy (Nyctaginaceae)	Baganvilas	Flower	Dyeing Blue
Commelina benghalensis L (Commelinaceae)	Kanshira	Flower	Painting blue
Nyctanthes arbor-tristis (Oleaceae)	Sheuli	Flower	Dyeing Yellow
Aegle marmelos Correa (Rutaceae)	Bel	Rind of the fruit	Printing (Reddish)
Anacardium occidentale L. (Anacardiaceae)	Kaju	Pericarp	Making ink (light red) and colour for net
Punica granatum L. (Punicaceae)	Dalim	Rind of the fruit	Dyeing sweets
Terminalia chebula Retz (Combretaceae)	Haritaki	Rind of the fruit	Dyeing sweets

3.36 EFFECT OF NATURAL BIO-FERTILIZER ON KENDU (*DIOSPYROS EXSCULPTA*) LEAVES

Abstract

The present work achieved great success to promote the production of kendu leaves having no adverse effect by foliar application of water-hyacinth extract.

Keywords : Bio-fertilizer, foliar spray, water hyacinth.

Introduction

Man has been using smoking materials since times immemorial for the pursuits of pleasure. Most of these materials have a stimulating effect due to presence of alkaloids. Dried, cured and fermented leaves of tobacco are enveloped by kendu (*Diospyros exsculpta*) leaves are called Biri which are extremely popular among tribals. Kendu trees grown well in red soil of forest in West rardh.

Kendu plant belongs to the family Ebenaceae. It is a medium sized tree. The stem bears ovate leaves. Commercially leaves are used in smoking industry. Moderate rainfall is required for the tree. The tree thrives best in a well-drained red soil. There is a great demand of increase in the size, weight, thickness and strength of the leaves. Generally the leaves of the tree mature within 4-5 months. The leaves turn greenish yellow during ripening. Leaves are harvested as they nipen. Harvested leaves are stacked or hung in curing barns. Curing takes within 10 days. The cured leaves are then sorted and similar grade leaves are tied into bundles for the purpose of marketing. The present paper aims at to increase the production of leaves both in quality and quantity by water-hyacinth.

Materials and Methods

To enhance leaf growth of kendu, leaf extract of water-hyacinth (*Eichhornia crassipes* solms) is sprayed thrice consecutively for 3 days at 15 days interval during morning.

Result and Discussion

Accumulation of Ca, Mg and N in leaves of water-hyacinth from waste water is a natural process. So, foliar spray of water-hyacinth extract on kendu trees induces luxuriant growth of the trees by absorbing different macro-elements which are essential for plants. In recent years application of water-hyacinth has been found in waste-water management (Goel *et al.* 1985). Aquatic plants especially water hyacinth have been widely used in waste water management for their effective capability of accumulating nutrients and other pollutants. Besides hormone like gibberelin which is also present in water-hyacinth may be a promoter for kendu leaves (Ghosh, 2004), brinjal etc.

4.
Glossary of Medical Terms

Abscess	A localized collection of pus in any part of the body. The result of disintegration of displacement of tissue.
Abortifacient	Anything used to cause or induce an abortion.
Ague	A popular name for malarial fever.
Alexeteric	Protective against infection, venom and poison.
Alixephermic	Antidotal.
Alternative	A drug which corrects disorder process of nutrition and restores the normal function of an organ or of the system.
Amenorrhoea	Suppression of menses not due to natural causes.
Anaemia	Condition in which there is a reduction in the number of circulating red blood per cu. mm, the amount of haemoglobin per 100 ml of blood.
Anodyne	A drug that relieves pains: an analgesic.
Anthelmintic	An agent that destroys parasitic intestinal worms.
Antihydrotic	A drug which checks sweating.
Antiperiodic	A drug which controls periodic attacks of disease; an anti-malarial drug.
Antispasmodic	A drug which prevents or cures colic, convulsions or spasmodic disorders.

Antipyretic	Reducing fever. An agent that reduces fever.
Antidote	A substance that neutralizes poisons or their effect.
Antisyphilitic	Curative or relieves syphilis.
Antiseptic	An agent capable of producing antisepsis.
Anticholinergic	Impending the impulses of cholinergic neurotransmitters. An agent that blocks parasympathetic nerve impulses.
Anthrax	Acute, infectious disease caused by *Bacillus anthracis*, usually attacking cattle, sheep, horses and goats. Man is infected in contact with animal hairs, hides or waste.
Aperient	A laxative or mild purgative.
Aphrodisiac	A drug which promotes sexual desire.
Apoplexy	Sudden loss of consciousness followed by paralysis caused by hemorrhage in the brain.
Aromatic	A drug having an agreeable odor.
Ascarides	A collection of fluid in the abdomen; abdominal dropsy.
Asthma	A disease of the bronchial tubes causing recurrent attacks of breathlessness and coughing.
Astringent	To bind fast, drawing together, constricting, binding. An agent that has a constricting or binding effect; i.e. one that checks hemorrhage or secretions by coagulation of proteins on a cell surface.
Benzaldehyde	A pharmaceutical flavouring agent derived from oil of better almond.
Billiousness	A symptom of a disordered condition of the liver causing constipation, headache, loss of appetite and vomiting of bile.
Black water fever	Billious remittent fever, a complication of malaria.

Belennorrhoea	Excessive mucous discharge, particularly from the urethra or vagina; gonorrhoea.
Blister	An accumulation of the fluid under the upper layer of the skin; a substance applied to the skin for raising a blister.
Boils	An infectious festering sore which ultimately may develop into an ulcer.
Bright's disease	An acute or chronic disease of kidneys.
Bronchitis	An inflammation of mucous membrane of the bronchial tubes or air passage; feverish cold with cough and sore chest.
Bruises	An injury with diffuse effusion into subcutaneous tissue and in which skin is discoloured but not broken.
Cardiac	Pertaining to heart.
Cardiotonic	Increasing tonicity of the heart. Various drugs, including digitalis, are cardiotonic.
Carminative	A drug which relieves flatulence or the feeling of over-fulness of the stomach.
Catarrh	Term formerly applied to inflammation of mucous membranes, especially of head and throat.
Cathartic	An active purgative producing bowel movements.
Cerebral	Pertaining to cerebrum; the brain.
Cholangogue	An agent that increases the flow of bile into the intestine.
Chorea	A nervous condition marked by involuntary muscular twitching of the limbs or facial muscles.
Colic	Spasm in any hollow or tubular soft organ accompanied by pain.
Cutaneous	Pertaining to skin.
Cystitis	Inflammation of bladder.
Dandruff	Normal exfoliation of the epidermis of the scalp in the form of dry white scales.

Demulcent	Stroking softly. An agent that will smooth the part or soften the skin to which applied.
Depressant	An agent that reduces functional activity.
Delirium	A state of mental confusion and excitement characterized by disorientation.
Depurative	Having cleaning properties; removing waste material from the body. Removal of waste material.
Deodorant	Perfumoral. An agent that masks or absorbs foul odours.
Diaphoretic	An agent that increase perspiration, such as camphor, opium or pilocarpine. Heat may also be included as such an agent. An agent that induces copious secretion of sweat.
Diarrhoea	Frequent passage of unformed watery bowel movement. It is a frequent symptom of gastrointestinal disturbance.
Diathesis	Constitutional predisposition to a certain disease, condition or group of diseases. Can be allergic, hemorrhagic or rheumatic disease.
Diabetes	A general term for diseases characterized by excessive urination. Usually refers to Diabetes mellitus.
Dropsy	Disease causing a watery fluid to collect in some cavity of the body.
Diuretic	An agent that increases the secretion of urine.
Dyspepsia	Imperfect or painful digestion. Not a disease in itself but symptomatic of other diseases or disorders. Indigestion.
Dysentery	An infectious disease, characterized by acute diarrhoea accompanied by gripping pains, the stools being chiefly of blood and mucous.

Dysuria	Painful and difficult urination.
Dysmenorrhoea	Pain in association with menstruation. One of the most frequent gynecologic disorders.
Eczema	Itching skin disease.
Elephantiasis	A disease of the skin and subcutaneous tissues causing hypertrophy of the affected parts. It may attack any part of the body but chiefly attacks the legs.
Emetic	An agent that produces vomiting.
Emollient	An agent that will soften and soothe the part when applied locally.
Emmenagogue	A substance that promotes or assists the flow of menstrual fluid.
Epilepsy	Nervous disease causing a person to fall unconscious (often with violent involuntary movement).
Expectorant	An agent that facilitates the removal of the secretion of the broncho-pulmonary mucous membranes. Classified as sedative or stimulating.
Febrifuge	An agent or drug used for reducing fever.
Febrile	Pertaining to fever, feverish.
Flatulence	Excessive gas in the stomach and intestines.
Freckles	Coloured spots, generally yellowish or brown, on the exposed parts on the skin.
Fructose	Levulose, fruit sugar.
Galactagogue	An agent that promotes the secretion and flow of milk; lactagogue.
Glect	A mucous discharge from the urethra in chronic gonorrhoea.
Gonorrhoea	A specific contagious, catarrhal inflammation of the genital mucous membranes of either sex.

Gravel	The development in the kidneys and urinary tract of tiny stone-like collection of uric acid, calcium oxalate or phosphates.
Granulation	Formation of granules or slate or condition of being granular.
Griping	A sharp colicy pain in the bowels due to presence of some irritating substance.
Haematuria	Passing of blood in the urine.
Haemorrhage	Bleeding, especially profuse, from any part of the body.
Haemorrhoids	Piles; a diseased condition of the blood vessels causing painful swellings in the region of the anus.
Hemicrania	Headache on only one side of the head; migraine.
Hepatitis	Inflammation of the liver.
Herpes	Creeping skin diseases.
Hernia	The protrusion or projection of an organ or a part of an organ through the wall of the cavity that normally contains it.
Hiccough/Hiccup	Spasmodic periodic closure of the glottis following spasmodic lowering the diaphragm, causing a short sharp, inspiratory cough.
Hydrophobia	Morbid fear of water. Common name of rabies resulting from bite of rabid animal.
Hysteria	A disease in which the patient who is physically healthy, suffers from imaginary diseases and has lost control over acts and feelings.
Hypnotic	Pertaining to sleep or hypnosis. An agent that induces sleep or that dulls the senses, such as chloral hydrate.
Induration	Area of hardened tissues.
Insomnia	An excessive amount of fibrin in the blood.

Intermittent	Suspending activity and intervals. Coming and going.
Ipacacauanha	It is a plant grown in Brazil. The dried root of this plant is known as Ipaca-cauanha. It is the source of emetine q.v.
Itch	An infectious disease of skin without specific lesions and marked by excessive itching.
Jaundice	A disease condition of the liver in which there is yellowish colouring of the tissues and urine with bile.
Laryngitis	Inflammation of larynx.
Leprosy	A chronic wasting disease caused by a germ which generally results in mutilations and deformities.
Leucoderma	A skin disease marked by skin losing its pigment wholly or partially.
Leucorrhoea	A vaginal discharge of a white fluid containing mucous and pus cells.
Lithiasis	The formation of calculi or stone in any part of the body.
Lithotriptic	A drug having the property of crushing a calculus or stone present in the urinary system.
Lithositic	Having the property of crushing a calculus in the bladder or urethra.
Lumbago	Rheumatism of the small of the back causing acute pain and stiffness.
Mania	Mental disorder characterized by excessive excitement.
Melancholia	A disorder of the mind marked by depression of spirits, mental sluggishness and apathy to ones surroundings.
Menorrhagia	Abnormally excessive menstruation.
Micturition	Urination.

Migraine	A nervous disorder marked by periodic attacks of headache; hemicrania.
Mucilaginous	Resembling mucilage, slimy; sticky.
Mumps	An infectious disease marked by inflammation of glands near the ear.
Mydriatic	A drug which dilates the pupil.
Narcotic	A drug which induces deep sleep or insensibility to pain.
Nausea	A feeling of sickness; inclination to vomit.
Nephritis	Inflammation of the kidneys.
Neuralgia	Severe sharp pain along the course of nerves.
Opthalmia	Inflammation of the eyes; conjunctivitis.
Otitis	Inflammation of the ear.
Otorrhoea	A purulent discharge from the ear.
Palsy	Temporary or permanent loss of sensation or loss of ability to move or to control movement.
Paralysis	Loss of motion in any fraction of any part of the body.
Paroxysms	A sudden, periodic attack or recurrence of symptoms of a disease. Sudden spasm or convulsion of any kind. Sudden emotional state, as of fear, grief or joy.
Pectoral	Pertaining to the chest; cough remedy; expectorant.
Photophobia	Unusual intolerance of light.
Phithisis	Affected with pulmonary tuberculosis.
Piles	An inflamed condition of veins in the rectal region; haemorrhoids.
Pleurisy	Inflammation of the membrane enclosing the lungs.
Pneumonia	Inflammation of the lungs.
Polyploidy	Condition in which the chromosome number is two or more times the normal haploid number found in gamets.

Poultice	A hot, moist mass of linseed, mustard or soap and oil between two pieces of muslin applied to the skin to relieve conjection or pain, to stimulate absorption of inflammatory products, and to act as a counter irritant.
Prolapse	A falling down of an organ from its normal position, especially its appearance at an opening.
Pruritus	Severe itching. May be a symptom of a disease process such as allergic response or be due to emotional factors.
Pulmonary	Pertaining to lungs.
Pungent	Sharp smell or taste.
Pyorrhoea	A purulent discharge from the gums.
Pyrosis	A burning sensation in the epigastric and sternal region with raising of acid liquid from stomach.
Rectum	Lower part of large intestine.
Refrigerant	An agent which relieves feverishness or produces a feeling of coolness.
Refringent	Refractive.
Rheumatism	A term used for pains in the muscles, joints and certain tissues; the disease takes various forms.
Resolvent	Promoting disappearance of inflammation.
Rubefacient	A mild counter-irritant; a drug that causes tingling reddening of the skin.
Schizophrenia	Mental disease that involves the disorganization of personality and characterized by delusions, hallucination, abnormal behaviour and retreat from reality.
Sciatica	A neuralgic pain at the back of the thigh caused by the inflammation of the sciatica nerve.

Scorbutic	Suffering from scurvy.
Scrofula	A disease chiefly of the young, marked by want of resisting power making the patient susceptible to tuberculosis especially of the glands, bones and joints, eczematous eruption, ulceration, glandular swelling, etc.
Sedative	A drug which has a calming or quieting effect on the patient and which reduces nervous excitement.
Sialagogue	A drug which promotes the secretion of saliva.
Soporific	A drug that induces sleep.
Spasmophile	A tendency to tetany and convulsion; almost always association with rickets.
Spasmodic	A convulsion. Concerning spasms.
Stomatitis	Inflammation of the mucous membrane of the mouth.
Stomachic	A drug which improves digestion and appetite.
Strangury	Painful and drop by drop discharge of urine.
Styptic	An agent which checks bleeding.
Sudroific	An agent that promotes perspiration; a diaphoretic.
Suppuration	The process of pus formation.
Syphilis	A serious chronic venereal disease.
Taenia	Tapeworms.
Tetanus	An infectious disease, marked by painful contractions in the muscles.
Tinea	A group of parasitic skin diseases, ringworm.
Tonsillitis	Inflammation of tonsil.
Trachoma	A contagious granular inflammation of the conjective.

Tuberculosis	A disease caused by bacillus; it may affect any part or organ of the body.
Ulcer	An open sore on the skin or on any mucous membrane.
Vermifuge	A drug which expels intestinal worms.
Vertego	Dizziness; giddiness.
Vesicant	A drug or agent that produces blisters.
Vulnerary	A drug which promotes healing of wounds.
Wart	A hypertrophy or growth on the skin.
Whooping cough	An acute infectious disease marked by recurring peculiar spasmodic attacks of coughing, each attack ending with a deep noisy intake of breath.

5.
REFERENCES

Abdi, Rupa Desai. 1993. Maldharis of Saurashtra-A Glimpse into their past and present. EXCEL Industries Pvt. Ltd., Bhavnagar.

Amiadi, S.R. 1977. Let's go to know our trees: Village trees. Sci. Today. 12 (1) : 31-35.

Anon, 1948-1976. The Wealth of India. A dictionary of Indian raw materials. CSIR, New Delhi, vol. I-XII.

Anonymous, 1983. Ethnobotany in India. BSI, Howrah.

Anonymous, 1986. Zilani Ankada-Ni-Kunhi Rooprekha (in Gujarati). Zila Parishad, Junagarh.

Anonymous, 1998. Medicinal plants. Their Bioactivity, Screening and Evaluation. Centre for Science and Technology of the Non-aligned and other Developing countries, New Delhi.

Arnon, D.I. 1949. Copper enzymes in isolated chloroplasts. Polyphenol oxidase in *Beta vulgaris.* Plant Physiol. 24 : 1-15.

Avicena, 1887. Qanon (Urdu Ed), Matba Munshi Naval Kishore, Lucknow. 2 : 114.

Babu, N. 1965. Observation on the toxicity of the seed of *Croton tiglium* on predatory weed fishes. Sci. Culti. 31 : 308-310.

Banerjee, D.K. 1974. Magico-religious beliefs about plants among some Adivasis of India. J. Mythic Soc. 65 (3) : 5-8.

Basak, S.K. 1997. Medicinal plants of Bankura (W.B.) and their uses. J Natl. Bot. Soc. 61.

Beatrice, H.K. 1975. Ethnobotany of the Hawaiians. Harold L. Lyon Arboretum, University of Hawaii.

Bell, J.N.B. and Clough, W.S. 1973. Depression of yield in rye grass exposed to sulphur dioxide. Nature (London). 241 : 47-49.

Bhatnagar, A.R., Vora, A.B. and Patel, T.S 1985. Measurement of dust fal on leaves in Ahmedabad and its effect on chloroplast. Indian J

Bhattacharya, G. 1996. Medico-ethno-botanical value of Saurashtra Weeds. In: Maheswari, J.K. (Ed). Ethnobotany in South Asia. J.E.T.B. Addl. Ser. 12, Sci. Pub., Jodhpur, India.

Bhattacharya, G. 2002. Ethnobotanical studies on some weeds of Gujarat. In Series Recent Progress in Medicinal plants, vol. 1. Ethnomedicine and pharmacognosy (Eds V.K. Singh, J.N. Govil and Gurdip Singh). Sci. Tech. Pub., Houston, Texas, USA, 33-40.

Bhuyan, B.R. 1967. Eradication of unwanted fish from ponds by using indigenous plant poisons. Sci. Cult. 33 : 82-83.

Billore, K.V. 1998. Interesting Folk remedies by the Lok Vaidyas of Rajasthan for Swasroga. Ethnobot. 10 : 12.

Biswas, A.K. and Ghosh, A.K. 1999. Regulation of senescence in various plants. Emkay Publications, Delhi.

Bole, P.V. and Pathak, J.M. 1988. Flora of Saurashtra, B.S.I., Howrah.

Brown, J.H. 1939. China berry poisoning in hogs. J. Am. Vet. Med. Assoc. 95 : 107-110.

Buttle, G.A., Darcy, P.F., Howard, E.M. and Kellet, D.N. 1957. Plethysmographic Measurement of Swelling in foot of Small Laboratory animals. Nature. 179 : 629.

Calesmisk, B. and Beutner, R. 1949. Inhibition of Hyaluronidase by Aromatic Compounds, Proc. Soc. Expt. Biol. Med. 72 : 629.

Chakraborty, D.P., Nandy, A.C. and Phillipose, M.T. 1972. *Berringtonia acutangula.* Gaertn as fish poison. Indian J. Exptl. Biol., 10 : 78 80.

Chattopadhyay, A. and Maji, M. 1975. Some medicinal trees of Bankura district of W. Bengal. Nagarjun. 19 (3) : 6.

Chattopadhyay, A. and Maji, M. 1976. Poor man's medicine. Man and Life. 2 : 126.

Chaudhuri, B. 1967. Magic Vs Medicine in a Tribal village. Adivasi. 9 : 3-9.

Chaudhuri Rai, H.N. and Pal, D.C. 1975. Notes on magico-religious belief about plants among Lodha of Midnapore, W. Bengal. Vanyajati. 23 (2-3) : 20-22.

Chaudhuri Rai, H.N., Saren, A.M. and Molla, H.A. 1982. Some less known use of plants from the tribal areas of Bankura district, W. Bengal. Indian Mus. Bull. 14 : 71-73.

Chettri, R., Rai, B. and Khawas, D.B. 1992. Certain medicinal plants in the folklore and folklife of Darjeeling (W.B.) and Sikkim Hills. J. Econ. Taxon. Bot. Addl. Ser. No. 10 : 395-398.

Chopra, R.N., Badhwar, R.L. and Ghosh, S. 1949. Poisonous Plants of India. Vol. 1, ICAR, New Delhi.

Cooke, T.H. 1958. The Flora of the Presidency of Bombay, B.S.I., Calcutta (reprinted). Vol. 1-111.

Cotton, M. 1996. Ethnobotany : Principles and Applications. John Wiley & Sons, Chichester, England.

Crittenden, I.D. and Read, D.J. 1978. The effects of air pollution on plant growth with special reference to sulphur dioxide II. Growth studies with *Lolium perenne* L. New Phytol. 80 : 49-62.

Dassanayake, M.D. 1978. Dye yielding plants of Sri Lanka-A List compiled by Andreas Nell. Phyta (Ceylon). 1 : 37-40.

De, S. 1968. Ethnobotany-A newer science in India. Sci. and Cult. 34 : 326-328.

Deb, D. and Banerjee, S. 1987. Muscle protein degradation of fishes exposed to mahua oil cake. Enviro. and Ecol. 5 (4) : 704-706.

Deb, D. and Malhotra, K.C. 1993. In : Peoples of India : Biocultural Dimensions (eds S.B. Roy and A. Ghosh). Inter India Publication. Delhi. 329-342.

Deb, D. and Malhotra, K.C. 1997. J. Human Ecol. 8 : 157-163.

Dutta, B.K. and Dutta, P.K. 2005. Potential of ethnomedicinal studies in Northeast India. An overview. IJTK. 4(1) : 7-14.

Faulks, P.J. 1958. Ethnobotany of the Pacific Northwest Indians. Econ. Bot. 19 : 378-382.

Gadgil, M. and Subhash Chandran, M.D. 1992. In Indigenous Vision : People of India. Attitudes to the Environment (ed G. Sen) . Sage, New Delhi, 183-187.

Gadgil, M. and Guha, R. 1995. Ecology and Equity, Routledge, London.

Gautam, R.D. 1994. Biological Pest Suppression. Westvill Publishing House, New Delhi, 221.

Gillam, S. 1989. The traditional healer as village health worker. J. Inst. Med. 67-76.

Ghosh, A.K. and Biswas, A.K. 1989. Monocarpic senescence of *Arachis hypogaea* : Nutrient withdrawal vs. Migration of senescence signal. Indian. J. Plant Physiol. 32 : 172-174.

Ghosh, A.K. and Biswas, A.K. 1991. Source – Sink relationship during ageing and senescence of *Solanum tuberosum*. Indian J. Plant Physiol. 34 : 25-29.

Ghosh, A., Maity, S. and Maity, M. 1996. Ethnomedicine in Bankura district, W. Bengal. J. Econ. Taxon. Bot. Addl. Series. 12 : 318-320.

Ghosh, A. 1999. Herbal veterinary medicine from the tribal areas of Bankura district, W. Bengal, J. Econ. Taxon. Bot. 23 : 557-560.

Ghosh, A. 2002. Herbal veterinary medicine from the tribal areas of Bankura and Midnapur district. In : Series Recent Progress in Medicinal Plants. Vol. 1 Ethnomedicine and Pharmacognosy (Eds : V.K. Singh, J.N. Govil and Gurdip Singh). Sci. Tech. Pub., USA pp. 233-237.

Ghosh, A. 2002. Ethnoveterinary medicines from the tribal areas of Bankura and Medinipur district, West Bengal, IJTK. 1(1) : 93-95.

Ghosh, A. 2003. Herbal veterinary medicine from the tribal areas of Medinipur and Bankura district, West Bengal. J. Econ. Taxon. Bot. 27(3) : 573-575.

Ghosh, A. 2003. Traditional vegetable dyes from Central West Bengal. J. Econ. Taxon. Bot. 27 : 825-826.

Ghosh, A. 2004. Natural biocides and biofertilizer, Natural Product Radiance. 3(3) : 173.

Ghosh, A. 2005. How one can survive for a long time in the earth. Ageing & Society. 15(3-4) : 75-81.

Ghosh, A. 2008. Ethnomedicine for Human and Veterinary Development, Daya Publishing House, Delhi.

Ghosh, R.B. and Das, D. 1999. A Preliminary census and systematic survey of anti-diabetic plants of Midnapore district, W. Bengal. Indian J. Econ. Taxon. Bot. 23 : 535.

Goel, P.K., Trivedy, R.K. and Vaidya, R.R. 1985. Accumulation of nutrients from waste-water by water-hyacinth, *Eichhornia crassipes*. Geobios. 12 : 115-119.

Harris, S. 1994. Honey for the treatment of superficial wounds: a case report and review. Primary Infections 2 (4) : 18.

Hooper, D. 1906. Some instances of vegetable pottery. J. Asia. Soc. 2 (3) : 65-67.

Horwitz, M.M. 1988. Restoring India's forest wealth. Natural Resource. 27 : 12.

Howe, D.K. and Woltz, S.S. 1981. Symptomatology and relative suceptibility of various ornamental plants to acute airborne sulphur dioxide exposure. Proc. Fla State Hort. Soc. 94 : 121-123.

Huang, K.C. 1999. Pharmacology of Chinese herbs (2nd edn) CRC Press, Boca Raton.

Hussain, A. 1992. Status report on medicinal plants for NAM countries. Centre for Science and technology of the Non-Aligned and other developing countries, New Delhi.

Jain, S.K. 1981. Glimpses of Indian Ethnobotany. Oxford and IBM. Publishing Co., New Delhi.

Kamble, K.D. 2003. Palm gur industry in India. IJTK. 2 : 137-147.

Karawja, M.S., Abdel, S.M., Wahab, M.M. El-Olemy and Farg, N.M. 1984. Diphenyleamine an antihyper glycaemic agent from onion & Tea. J. Nat. Pr. 47 : 775-780.

Kreig, M.S. 1964. Green medicine : The search for plants that heal, Rand Mcnally, New York and London.

Lal, K.S. 1980. The Mughal Harem. Aditya Prakashan, New Delhi.

Latif, A., Tarif, M. Afatq, S.H. and Asif, M. 1981. Effect of *Cardiospermum halicacabum* in Russell's Viper induced envenomation. XIV Annual Conference IPS, Bombay, Souvenir. 32 : 46.

Lezy, 1964. Introduction of cattle by capsules of the field poppy. Rec. Med. Veterinary. 122 : 23-24.

McClure, F.A. 1927. Note on a Chinese vegetable dye (*Dioscorea rhipogonoides*). Lingnaam Agric Rev. 4 : 1-5.

McIntyre, M. 1994. European herbal remedies. Paper presented at the meeting towards the safer use of Traditional Remedies, Royal Botanical Garden, Kew.

Mondal, S.R. 1989. The Eunuch. Some observations. J. Indian Anthro. Soc. Vol. 24(3).

Mondal, M.K. and Chauhan, J.P.S. 2000. A survey of ethnoveterinary medicine practices in West Bengal. Indian J. Vet. Med. 20(1) : 90-91.

Moorby, J. and Milthrope, P.L. 1975. Potato. In : Crop physiology (L.T. Evan, ed.) Cambridge University Press. 225-257.

Mukherjee, J.B. 1978. Castration as a means for induction into Hijrah Community (Eunuch) in India in 8th International Forensic Conference held in USA.

Mukherjee, A. and Namhata, D. 1988. Herbal veterinary medicine as practised by the tribals of Bankura district, W. Bengal. J. Bengal Natural Hist. Soc. 7 : 69-71.

Mutalik, S. 1991. Comment : The use of honey and sugar for the treatment of ulcer in leprosy. Lepr. Rev. 62 (2) : 228.

Nadkarni, K.M. 1954. Indian Materia Medica, vol. 1 (3rd ed). Popular Book Depot. Bombay, 271-272.

Naik, T.B. 1951. Aboriginals of Gujarat. The Eastern Anthropologist. 4 : 42-50.

Namhata, D. and Mukherjee A. 1988. Ethnomedicine in Bankura district, W. Bengal. Indian J. Applied and Pure Biol. 2 : 53-55.

Namhata, D. and Ghosh A. 1993. Herbal folk medicine of Bankura district, W. Bengal. Geobios new reports. 12 : 94-96.

Naskar, K.R. 1986. Trends off aquatic weed management through proper utilization. Proc. Nat. Sem. Recent Trends in Plant Sci. Res. Santiniketan.

Naskar, K.R. and Guha Bakshi, D.N. 1987. Mangrove swamps of the Sunderban. An Ecological Perspective, Naya Prakash, Calcutta.

Nayak R.H. 1965. Tree cult in Karnataka. Folklore. 6 : 238-240.

Neto, E.M.C. 1999. Traditional use and sale of animals as medicines in Feira de Santana City, Bahia, Brazil. Indigenous Knowledge Development Monitor. 7 : 15.

Pal, D.C. and Jain, S.K. 1989. Notes on Lodha medicine in Medinipur district, West Bengal. Economic Bot. 43(4) : 464-470.

Prakash Ved. 1998. Indian Medicinal Plants : Current Status-1. Ethnobotany. 10 : 112-121.

Pramanik, P. 1956. Santal superstition regarding ploughing. Vanyajati. 4 (4) : 154.

Puri, H.S. 1970. Drugs of animal origin used in Indian system of medicine. Nagarjun. 13 : 21-23.

Rao, D.N. 1972. *Mangifera indica* L. A bioindication of air pollution in the tropics. Proc. 22nd Internat. Geograph. Congr. Montreal. 292-293.

Ray, C.R. 1922. On tree cults in the district of Midnapur in south-western Bengal. Man in India. 2 : 242-264.

Roy, S.N. 1925. Some popular superstitions of Orissa. Ibid. 5 : 210-234.

Roy, Shankarnath. 1977. Bharater Sadhak (Vol. 1-8), Karuna Prakashani, Kolkata.

Sahay, K.N. 1964. Tree-cult in tribal culture. Folklore. 5 (7) : 241-257.

Santapau , H. and Janardhanan , K.P. 1967. The Flora of Saurashtra-Check List. Bull B.S.I. : 8 (Suppl. 1) : 1-58.

Satyavati, G.V. 1991. Guggulipid : A promising hypolipidaemic agent from gum guggul *Comiphora wightii* in Economic and Medicinal plant Research vol. 5. Plants and Traditional medicine (eds Wagener, H. and N.R.) Farnsworth academic Press, London.

Sen, S. and Batra, A. 1997. Household remedies from Rajasthan. Jeevaniya. 8 (1) : 14.

Sen, S. and Batra, A. 1998. Household remedies for constipation and abdominal pain. Jeevaniya. 8 : 14-15.

Sethi, R. 2004. Health Drinks. Natural Product Radiance. 3(1) : 16-18.

Shah, G.L. 1978. Flora of Gujarat state. Sardar Patel University, Gujarat.

Shanmughavel, P. 2004., Cultivation potential of culinary bamboos in S. India. Natural Product Radiance. 3 : 237-238.

Sharma, P.V. 2000. Charaka Samhita, Sutra Sthana (Chapter 6), 6th edition. Chaukhamba Orientalia, Varanasi, India.

Shawl, H.Y., Tripathi, L. and Bhattacharya, S. 2004. Antidiabetic plants used by tribals in M.P. Natural Product Radiance. 3(6) : 427.

Shukla, S.D., Modi, N.T. and Deshmenkar, B.S. 1973. Some pharmacological action of alkaloidal fraction from *Cardiospermum halicacabum*. Indian J. Pharma. 35(2) 40.

Singh, V.K., Govil, J.N. and Singh, G. 2002. Recent progress in medicinal plants. Sci. Tech. Publishing Ltd., USA.

Sinha, S.P. 1976. Tribal Gujarat Directorate of Information. Govt. Press, Gandhinagar.

Sreedevi, B. 1994. Indigenous technical knowledge n farming agricultural research. Kisan World. 21 (12) : 11.

Sukh Dev 1997. Ethno therapeutics and modern drug development : the potential of Ayurveda. Current Sci. 73 : 909.

Suryanarayan, M.C. 1986. Honey bee – flower relationship. Bull. Bot. Surv. India. 28 : 55.

Tyler, E., Varro, Lynn R., Brady, James Robbers. 1981. Pharmacognosy, 8th Edition, USA.

Vedavarthy, S. 2004. *Decalepis hamiltonii* Wight 8 arm. An endangered source of indigenous health drink. Natural Product Radiance. 3(1) : 22-23.

Vedavartty, S. Mrudula, V. and Sudhakar, A. 1997. Tribal medicine of Chittor district (A.P.). Herbal Folklore Research Centre, Tirupati, India, 165.

Viswanathan, N. and Joshi, B.N. 1983. Toxic constituents of some Indian plants. Current Sci. 52 (1) : 1-8.

Waite, M.B. et al. 1925. Yearbki, U.S. Dep. Agric. 453.

Watt, G. 1989. The dictionary of the economic products of India. Vol. 1-6, DEP, Calcutta.

Winter, C.A. Risely, E.A., Nuss, G.W. 1962. Carragenin induced oedema in Hind paw of rats. An assay fro anti-inflammatory drugs. Proc. Soc. Exp. Biol. Med. III 544-547.

6.
List of Publications by Author in East & West

1. J. Agronomy & Crop Science 162, 342-346 (1989) 1989. Paul Parey Scientific Publishers. Berlin and Hamburg ISSN 0931-2250.

   Monocarpic Senescence in Relation to Yield of *Sesamum indicum.* During Source – Sink Alteration A.K.Biswas and A.K.Ghosh.

2. Indian J. Plant Physiol., Vol. XXXII, No. 2. pp. 172-174 (June, 1989)

   Monocarpic Senescence of *Arachis hypogaea* : Nutrient Withdrawal vs. Migration of Senescence Signal. A.K. Ghosh and A.K. Biswas.

3. Indian Journal of Experimental Biology. Vol. 28. May 1990. pp. 492-493.

   Whole plant senescence in *Ophioglossum vulgatum* : Influence of sporangiferous spike, kinetin and abscisic acid.

   Arun Kumar Biswas*, Ashis Kumar Ghosh, Swapan Kumar Mandal & Udayan Sarkar.

4. Indian J. Plant physiol., Vol. XXXIV No. 1. pp. 25-29 (March, 1991).

   Source- Sink Relationship during Ageing and Senescence of *Solanum tuberosum* L.

   Ashis Kumar Ghosh and Arun Kumar Biswas.

5. Agri. Biol. Res. 7 (2) 132-138 (1991).

   Source- Sink Relationship during Monocarpic Senescence of *Raphanus sativus.*

   Ashis Kumar Ghosh and Arun Kumar Biswas.

6. Indian Journal of Forestry Vol. 16 (3) 201-203. 1993.

Intact Leaf Senescence in Some Polycarpic Forest Trees in Relation to their Reproductive Development.

Ashis Kumar Ghosh and Arun Kumar Biswas.

7. Geobios new Reports 12 : 94-96, 1993.

Herbal Folk Medicines of Bankura District, West Bengal.

D. Namhata and A. Ghosh.

8. Geobios new Reports 12 : 96-99, 1993.

Effect of Polluted Air of Durgapur on the Vegetation.

Ashis Kumar Ghosh.

9. REVISTA Biol (Lisboa) 15 : 63-68, 1994

Monocarpic Senescence in *Amaranthus tricolor* L. : Dual Action of Kinetin.

Ashis Kumar Ghosh and Arun Kumar Biswas.

10. Indian Journal of Experimental Biology. Vol. 32. November 1994, pp., 807-811.

Correlative senescence in *Tagetes patula* and *Chrysanthemum coronarium* during reproductive development.

Ashis Kumar Ghosh & Arun Kumar Biswas.

11. J. Agronomy & Crop Science 175-202 (1995)

(c) 1995 Blackwell Wissenshafts Verlag, Berlin ISSN 0931-2250.

Regulation of Correlative Senescence in *Arachis hypogaea* L. by Source-Sink Alteration through Physical and Hormonal Means A.K. Ghosh and A.K. Biswas.

12. J. Econ. Taxon. Bot. Additional Series, 12

Scientific Publishers, Jodhpur (India), 1996, pp. 318-320

Ethnomedicine in Bankura district, West Bengal. Ashis Ghosh, Sarathi Maity & Malaty Maity.

13. Fifth West Bengal State Science Congress held at University of North Bengal, March 21-23, 1998 Herbal Veterinary medicine from the Tribal Area of Bankura District, West Bengal. A.K. Ghosh.

14. The Botanica, Vol 42, pp. 20-23 (1993) Magazine of Delhi Univ. Botanical Society, Inhibitory Effect of Fishes Exposed to some Cultivated, Wild plants from the Bangiya West Rardh (Bankura district). Ashis Ghosh.

15. Pak. J. Sci. Ind. Res. 2002, 45 (3) 212.

    Mechanism of Monocarpic Senescence of *Trichosanthes dioica* Roxb (Cucurbitaceae). Ashis Ghosh.

16. Ghosh Ashis Kumar (2002). "Herbal Veterinary Medicine from the Tribal Areas of Bankura and Midnapur Disrtict". In : Recent Progress in Medicinal Plants. Vol. 1- Ethnomedicine & Pharmacognosy (Eds : V.K.Singh, J.N. Govil & Gurdip Singh). SCI TECH Pub., USA, 233-237.

17. Indian Journal of Traditional knowledge, Vol. I (i) October, 2002, pp. 93-95. Ethnoveterinary medicines from the tribal areas of Bankura and Medinipur districts, West Bengal. Ashis Ghosh.

18. J. Econ. Taxon. Bot. Vol. 27 No. 3 (2003), Scientific Publishers (India), pp. 573-575. Herbal veterinary medicine from the tribal areas of Midnapur & Bankura district, West Bengal, Ashis Kumar Ghosh.

19. J. Econ. Taxon. Bot. Vol. 27 No. 4 (2003), Scientific Publishers (India), pp. 825-826. Traditional Vegetable dyes from central West Bengal. Ashis Ghosh.

20. Indian Journal of Traditional knowledge, Vol. 2 (4) October, 2003, pp. 393-396. Herbal folk remedies of Bankura and Medinipur districts, West Bengal. Ashis Ghosh.

21. Natural Product Radiance Vol 3(2) p-91 March-April 2004. Plant and Clay dyes used by weavers and potters in West Bengal. Ashis Ghosh.

22. Natural Product Radiance Vol 3(3) p-170 May-June 2004. Apiphilic plants in agro-forestry. Ashis Ghosh.

23. Natural Product Radiance Vol 3(3) p-173 May-June 2004. Natural biocides and biofertilizers. Ashis Ghosh.

24. Natural Product Rudiance Vol 3(6) p-426 Nov-Dec 2004. Home made baby food. Ashis Ghosh & Jhuma Karan

25. Pak. J. Sci. Ind. Res. 2005, 48 (1) 55-56 Mechanism of Monocarpic Senescence of *Momordica dioica* : Source-Sink Regulation by Reproductive Organs. Ashis Ghosh.

26. Non-wood News, No.12 p-30 March 2005, Rome, Italy. Plant and Clay Dyes. Ashis Ghosh.

27. Ageing & Society : The Indian Jounal of Gerontology, Vol. 15 No. 3 & 4 : 75-81, 2005.

    How one can survive for a long time in the earth. Ashis Ghosh.

28. Ghosh Ashis (2006). Identification of Veterinary Medicinal plants through participatory approach. In : Advances in Medicinal Plants (Eds. N.D. Prajapati, T. Prajapati & S. Jajpura) Vol. 2, pp. 97-105. Asian Medicinal Plants & Health Care Trust, Jodhpur, Rajasthan.

29. Natural Product Radiance, Vol. 5(4) : 260. July-Aug. 2006. Healthy Pan-drink. Ashis Ghosh.

30. J. Econ. Taxon. Bot. Vol. 30 (Suppl.), pp. 233-238, 2006. Medicinal plants used for treatment of diabetes by the tribals of Bankura, Purulia and Medinipur of West Bengal. Ashis Ghosh.

31. Natural Product Radiance. Ethnobotanical Survey in West Rardh for natural health care and green belt movement. Ashis Ghosh.

Books

1. Regulation of Senescence in Various Plants by A.K. Biswas & A.K. Ghosh, Emkay Publications, Delhi.
2. Paribesh O Udvid, Ashis Ghosh.
3. Prani Rahasya, Ashis Ghosh.
4. Ethnobiology : Therapeutics & Natural Resources, Ashis Ghosh, Daya Publishing House, New Delhi.
5. Ethnomedicine for Human and Veterinary Development, Ashis Ghosh, Daya Publishing House, New Delhi.

7. INDEX

7.1 PLANT INDEX

A

Abelmoschus esculentus 19
Abroma augusta 10
Abrus precatorius 74, 78, 161
Acacia arabica 35
Acacia auriculiformis 112
Acacia catechu 63, 66, 136, 144, 157
A. concinna 10, 74
A. leucophloea 9
Acacia nilotica 168
Acanthus ilicifolius 4
Achras zapota 20, 40, 46, 135
Achyranthes aspera 39, 72, 161
Acrua lanata 12
Adansonia digitata 16
Adina cordifolia 39, 46, 105
Aegiceras corniculatum 112
Aegle marmelos 19, 35, 129, 135, 144, 157, 161, 168
Aerva lanata 162
Alangium salviifolium 12, 17
Alanthus exceisa 16
Albizia lebbeck 39, 45, 66
Allium cepa 35, 129, 135, 157, 160
A. sativum 137, 142, 153
Alocasia indica 39, 45
Aloe indica 28
vera 19
Alstonia scholaris 78
Amaranthus spinosus 40, 45
A. tricolor 35, 126
Anacardium occidentale 78, 168
Annona reticulata 142
A. squamosa 72
Anthocephalus cadamba 39, 46
Arachis hypogaea 112, 129, 159
Argemone mexicana 10, 78, 160
Aristolochia indica 5, 9, 11, 143, 162
Artemisia vulgaris 10
Artocarpus heterophyllus 9, 19, 138
Averrhoa carambola 16
Avicennia alba 112
A. officinalis 4, 112
A. marina 112
Azadirachla indica 16, 25, 35, 55, 74, 105, 112, 136, 153, 166

B

Bambusa arundinacea 164
B. bambos 164
B. tulda 164
B. vulgaris 12, 164

Barringtonia acutangula 74
Bauhinia purpurea 168
Benincasa hispida 159
Biophytum sensitivum 16
Boerhavia diffusa 16
  B. repens 4
Bombax ceiba 159
Borassus flabellifer 28, 39, 95, 119
Bougainvillea glabra 158, 168
Brassica campestris 11, 22, 112, 131
Bruguiera cylindrica 3
  B. gymnorrhiza 3
  B. parviflora 3, 5
Bryophyllum calycinum 35, 39
Buchanania lanzan 105
Butea monospherma 64, 66, 152
  B. superba 168

C

Caesalpinia crista 10, 39, 45
Cajanus cajan 25, 116, 137, 152
Calotropis gigantea 76, 78
  C. procera 35, 162
Camellia sinensis 64, 66, 135, 159, 166
Cannabis sativa 72, 79, 157
Capsicum frutescens 23, 142, 159
Cardiosphermum halicacabum 41-44
Carica papaya 13, 19, 24, 153, 157
Casearia elliptica 77, 78
Cassia fistula 17, 138, 168
Carthamus tinctorius 63
Casuarina equisetifolia 168
Catharanthus roseus 79
  C. pusillus 79
Centella asiatica 16
Cerbera manghas 4
Ceriops decandra 3, 66, 113
  C. tagal 5, 66, 113, 168
Cephalandra indica 39, 157
Cicer arietinum 159
Cinnamomum zeylanica 11, 135
Cissus quadrangularis 40
Citrus acida 22
  C. aurantifolia 19, 23, 26, 66, 129
C. limon 28
  C. maxima 19
  C. reticulata 19, 166
  C. sinensis 35
Clemalis gouriana 75, 79
  C. hedysanfolia 16
Cleistanthus collinus 79
Cleome viscosum 76
Clerodendrum infortunatum 76
Clitoria ternatea 12
Cocos nucifera 36, 137, 158, 159
Commelina benghalensis 168
Corchorus capsularis 144
Coriandrum sativum 131, 161
Costus mexicanus 135
Croton banplandianum 68-70
  C. tiglium 73
Cucumis sativus 9, 19, 22, 138
Cuminum cyminum 137, 154
Curculigo orchioides 17
Curcuma longa 24, 39, 40, 45, 46, 64, 153, 162, 166
Cuscuta reflexa 11, 77, 79
Cynodon dactylon 36

D

Datura metel 11, 40, 79, 116, 135, 160
D. stramonium 80, 117, 153
Daucas carota 64, 66, 138
Derris indica 75
D. scandens 75
D. trifoliata 4
Desmodium gyrans 12
Diospyros exsculpta 137, 162, 169
D. peregrina 66
Drosera burmannii 8
Dryopteris filix-mas 116

E

Eclipta alba 16, 17
Eichhornia crassipes 169
Emblica officinalis 20, 137, 166
Enicostema hyssopifolium 16
E. globulus 112
Eucalyptus citrodora 112
Eupatorium ayapana 11
E. odoratum 58-70
Euphorbia antiquorum 73, 80
E. hirta 73
E. nerifolia 80
E. tirucalli 28, 80

F

Feronia elephantum 25
Ferula asafoetida 26, 129, 143
Ficus benghalensis 36, 105, 138
F. carica 152
F. religiosa 105, 159
Finlaysonia obovata 4
Flacourlia indica 10
Foeniculum vulgare 162

G

Garcinia mangostana 9
Gardenia aqualis 161
Ginkgo biloba 33
Gloriosa superba 80
Glycine max 137
Glycorrhiza glabra 166
Gossypium herbaceum 136
Gymnema sylvestre 13

H

Hemidesmus indicus 10, 11, 166
Heritiera fomes 5
Hibiscus esculentus 136
Holarrhena antidysenterica 17, 138
Hybanthus enneaspermus 17
Hydrocarpus kurzii 137

I

Ipomoea aquatica 40, 46
I. batatus 157
I. bona-nox 68
I. digitata 138
I. fistulosa 29, 142
I. paniculata 12

J

Jatropha curcas 80, 145
J. gossypifolia 73, 80
J. hastata 152
Juglans regia 20

L

Lactuca sativa 127, 129, 137
Lantana camara 68-70, 116, 117
Laportea crenulata 81
Lathyrus aphaca 81

Lawsonia alba 64
L. inermis 36, 160
Leucas aspera 17
L. urticaefolia 17
Litchi chinensis 131
Litsea polyantha 152
Ludwigia perennis 16
Lycopersicon esculentum 9, 166

M

Machilus bombycina 152
Madhuca indica 76, 105, 112, 168
Malachra capitata 11
Malus sylvestris 20
Mangifera indica 112, 117, 131, 135
Martynia diandra 20
Maytenus emarginata 16, 17
Melia azedarach 74
M. azadirachta 74
Mentha piperila 166
Metroxylon sagus 144
Michelia champaca 95
Mikania scandens 68, 69
Mimosa pudica 12, 16, 162
Mimusops elengi 66
Momordica charantia 23, 138
Morinda citrifolia 168
Moringa oleifera 4, 36, 40, 138, 160, 166
Morus alba 20, 152
M. indica 152
M. levigata 152
M. serrata 152
Mucuna gigantea 4
M. pruriens 81
Murraya koenigii 158
Musa paradisiaca 19, 126, 136
M. sapientum 19, 20, 22

N

Nauclea orientalis 75
Nelumbo nucifera 36, 113
Nerium indicum 39, 81
Nicotiana rustica 76
N. tabacum 76, 117
Nigella sativa 40
Nyctanthes arbor-tristis 63, 168
Nymphaea alba 168
N. rubra 113
Nypa fruticans 3, 119

O

Ocimum canum 17
O. sanctum 166
Opuntia dillenii 72
Oroxylum indicum 161
Oxalis corniculata 80
Oxystelma secamone 17

P

Pachyrhizus angulatus 29, 142
Paederia foetida 40, 126, 143
Papaver somniferum 40, 46, 73, 81
Parthenium hysterophorus 81
Pedilanthus tithymaloides 81
Phaseolus aureus 23, 121
P. mungo 11, 144
Phoenix dactylifera 119, 166
P. paludosa 3
Phyllanthus niruri 166
P. urinaria 157

Piper betle 22, 36, 153, 154, 166
P. longum 154
P. nigrum 13, 40, 45
Plantago ovata 158
Plumbago zeylanica 82
Plumeria acutifolia 153
Plumeria rubra 82
Polygonum hydropiper 75
Porteresia coarctata 3
Portulaca oleracea 136, 157
Prunus persica 168
Psidium guajava 36, 157, 166
Pterocarpus marsupium 16, 105, 168
P. santalinus 62
Pterospermum suberifolium 76
Punica granatum 19, 62, 66, 135, 166, 168

R

Ranunculus scleratus 82
Rauvolfia canescens 12, 13
R. serpentina 161
Rhizophora apiculata 3, 66, 135
R. mucronata 3, 66
R. stylosa 66
Rhynchosia minima 17, 75
Ricinus communis 28, 82, 153, 160
Rivea hypocrateriformis 55
Rubia cordifolia 168
R. tinctorium 64

S

Saccharum officinarum 22, 144, 147
Salicornia brachiata 82
Santalum album 138
Sarcostemma acidum 161
Semecarpus anacardium 17, 82
Sesamum indicum 11, 113, 126, 131
Sesbania grandiflora 126
S. sesban 74
Shorea robusta 105, 112, 137, 152, 162
Sida cordifolia 9
Smilax indica 10
Smithia germiniflora 160
Solanum melongena 40
S. surathense 9
S. tuberosum 149
S. xanthocarpum 12
Spinacea oleracea 129
Spirulina platensis 166
Spondias pinnata 40, 46
Steudnera virosa 82
Stevia rebaudiana 134
Streblus asper 52, 95
Strychnos nux-vomica 77, 82
Swietenia mahogoni 66
Syzygium aromaticum 22
S. cumini 19, 131
S. jambos 37, 131, 135

T

Tagetes erecta 63, 67, 117
T. patula 63, 67, 111
Tamarindus indica 22, 37, 40, 45, 72
Tephrosia candida 75
Terminalia arjuna 13, 138, 152, 168
T. catappa 137
T. chebula 22, 64, 146, 166, 168
T. tomentosa 105, 152

Thevetia nerifolia 142
Tinospora cordifolia 17, 136, 143, 161
Trachyspermum amni 22, 23
Tragia involucrata 83
Trianthema portulacastrum 17
Trichosanthes anguina 37
Trichosanthes dioica 37, 77, 83, 142
Tridax procumbens 68-70
Trigonella corniculata 136
  T. foenum-graecum 16, 144, 158, 166
Triticum aestivum 121, 129
Tylophora indica 83

V

Vinca rosea 136, 143, 153
Vitex negundo 9, 37, 40, 143, 159
Vitis vinifera 166

W

Withania somnifera 136

X

Xanthium strumarium 68, 69
Xylocarpus granatum 3, 5

Z

Zea mays zanthoxylum aromatum 83
Zingiber officinale 23, 153, 154, 166
Ziziphus jujuba 13, 67, 152
  Z. mauritiana 131, 160
  Z. oenoplia 67

7.2 ANIMAL INDEX

A

Acridotheres tristis 93, 99, 108
  A. fuscus 99, 108
  A. ginginianus 50
Anthus novaeseelandiae 108
Apus affinis 108
Ardeola grayii 97
Athena brama 109

B

Bos indicus 52, 140
Bubalus sp. 51
Buthus meroccanus 56

C

Camefus dromedarius 52
Canis aureus 53, 59
  C. familiaris 52
Capra indicus 52
Caprimulgus asialicus 108
Catla catla 93
Ceryle rudis 108
Centropus sinensis 108
Cervus unicolor 55
Chunna striatus 58
Cisticola juncidis 109
Columba livia 51, 108

Copsychus saularis 100, 109
Coraclus benghalensis 108
Corvus splendens 50, 100, 109
  C. macrorhynchos 109
Cuculus variuos 108
Cyanopterus sphinx 51
Cypsiurus parvus 108

D

Dendrocitta vagabunda 109
Dicrurus adsimilis 109
Dinoplum benghalensis 98, 109

E

Equus heminous Khur 53
Eryz johni 59
Eudynamys scolopacea 108

F

Francolinus pondicerianus 108
  F. francolinus 108

G

Gallus domesticus 50
Gyps bengalensis 108

H

Halcyon smyrnensis 98, 108
Hippocampus cuda 58
Hirudinaria granulosa 56
Hyaena hyaena 53, 59
Hystrix indica 54, 160

L

Lepus nigricollis ruficodatum 54, 59
Lissemys punctatus 55

M

Megalaima haemacephala 98, 109
Milvus migrans 108
Murex brandaries 62
Mututa victor 57

N

Nectarinia zeylonica 109

O

Oriolous xanthomus 109
Orthotomous sutorius 100, 109

P

Panthera tigris 54
Paraechinus microlapus 53
Paradoxurus hemaphroditus 59
Passer domesticus 50, 109
Pavo cristatus 51, 59
Pediculus humanus 56
Periplanata americana 56
Phalacrocorax niger 97
Pheritima posthuma 56
Ploceus phillipinus 109
Potaman koolovensis 57
Presbytis entellus 59
Pseudibis papillosa 50
Psittacula krameri 109
  P. eupatria 109
Pteropus giganteus 59
Pycnonotus cafer 100, 109
  P. jocosus 99

R

Ratus ratus 54

S

Streptopelia chinensis 108
S. decaocto 108
Sturnus contra 99, 109
Sus scrofa 59

T

Treron phoenicoptera 108
Turdoides striatus 109
Turnix susciator 108
Tyto alba 109

U

Uca pughex 57
Upupa epops 108
Uromastix hardwickii 55

V

Varanus bengalensis 56, 59
Viviparus benghalensis 58

Z

Zygaena blochii 58

7.3 DISEASE INDEX

A

Abortifacient 82
Abscess 123
Accumulation of fat 9
Acidosis 127
Acidity 35, 88
Acne 11, 96
Ague 126
Alcoholic intoxication 59
Allergy 78, 88
Alopecia 10, 11, 46, 54
Anaemia 19, 88, 123, 128
Antidiabetic 14, 16, 52, 53
Antidiarrhoeal 120
Antifertility 10, 64
Anti-inflammatory 41, 42, 44
Antipyretic 14, 16
Antispasmodic 42
Aphrodisiac 58
Apoptosis 140
Appendicitis 88
Arthritis 41, 42, 50, 52, 53, 146
Asthma 4, 12, 50, 53, 56, 57, 61, 129

B

Back pain 59, 90
Bad smell 154
Baldness 54
Biliousness 20
Blindness 71, 73, 78, 79
Blister 78, 79, 81, 82
Blood cancer 136
Blood coagulation 40
Blood dysentery 10, 20, 40, 46
Blood purifier 4
Blood sugar 10, 13, 19, 27, 36, 127, 138
Blood vomiting 9

Boils 10, 51, 123
Bone fracture 40, 50, 52, 160
Brighty skin 53
Burning sensation 80, 81, 83
Burns 124

C

Cancer 9, 58, 86, 124, 140, 142
Candidiasis 36
Carbuncle 123
Cast-off skin 40
Castrated 30, 31
Cataract 5, 36, 58, 90
Chicken pox 36, 37
Cholera 123, 129, 162
Cholesterol 19, 37, 134, 138
Cold and cough 4, 56, 59, 66, 87, 90, 124
Colic pain 39, 46, 56, 123, 124
Colities 128
Coma 80
Conjunctivitis 55, 58
Constipation 13, 19, 20, 51, 88, 124, 126, 157, 159
Contraceptive 10, 20, 161
Convulsion 90
Crying 90
Cut wound 28, 35, 36

D

Dandruff 10, 123, 160
Defoliation 70
Depression 19, 35, 61, 79
Dermatitis 80, 81
Diabetes 13, 20, 33, 35, 86, 90, 132-139, 141, 158
Diarrhoea 4, 10, 11, 13, 19, 78, 157, 158, 162
Dimness of vision 12, 157
Diptheria 154
Diuretic 128
Dog bite 28, 161
Dropsy 78
Drowsy 81
Dysentery 19, 28, 36, 59, 157
Dyspepsia 19, 20, 22, 124

E

Earache 51, 57
Eczema 81, 82, 129
Emetic 80, 81
Enlargement of liver 12
Enuresis 37
Epilepsy 59, 88
Eruption 81

F

Fatal 82
Fatigue 88, 126
Fermentation 132, 169
Fertilization 124
Fever 51, 81, 87, 124, 160
Flatulence 13, 40, 45
Food poison 9, 11
Foot and mouth disease 57, 59, 160

G

Gastric ulcer 58, 123
Gout 55, 56, 124
Grooming hair 5

H

Habitual abortion 36
Haemorrhage 5
Hair greying 10
Hallucination 79, 80
Headache 61, 87
Heart trouble 61
Hernia 88
High pressure 19, 35, 36, 56, 82, 88, 127, 129, 161
Histiria 124, 129
Hydrophobia 12, 161
Hydrotheraphy 124
Hydrotuberation 124

I

Icteric 14, 17
Immunity 31
Impotence 10, 124
Indigestion 35, 124
Infertility 11, 12
Inflammation 123, 124
Insect bite 39, 162
Insecticide 74-76, 142, 143
Insomnia 19, 31, 36, 81, 86, 124, 129
Irritant 80, 81
Itching 81, 83, 133

J

Jaundice 19, 88, 157, 160
Joint sprain 58

K

Kidney stone 11, 58

L

Lactogenic 4
Late blight 149, 150
Lathyrism 81
Laxative 20, 127, 145
Leech sucking 45
Leprosy 4
Leucoderma 59
Leukaemia 157
Lice infestation 123
Liver disorder 19
Locked jaw 29
Low pressure 12

M

Madness 79
Malaria 87
Mating call 51
Measles 129
Menopause 37
Migraine 88, 123
Miliaria 11
Milk production 4, 41, 45
Miscarriage 10
Mumps 56

N

Narcotic 80, 81
Necrosis 70
Nephritis 129
Nervous disorder 50, 129
Neurosis 124
Night blindness 157
Night wetting 10

O

Oozes out of milk 28
Obesity 32, 128, 129
Oedema 41

P

Pain killer 56
Paralysis 50, 51, 129
Parturition 39
Piles 5, 28, 59, 88, 145
Piscicides 4, 70-77
Pneumonia 52, 146
Pollutant sink 6, 68
Pox 12, 129, 162
Pregnancy 140
Protein deficiency 4
Pruritis 55
Purgative 80, 82

R

Rabies 54
Rat bite 39, 45
Recurrent breeding 158
Rheumatism 4, 41, 42, 159

S

Sciatica pain 53, 56
Sea sickness 56 124
Seedling rot 158
Senescence 20, 23, 29-33, 57, 67, 126-128, 140, 154, 165
Shoulder wound 39, 45
Sinus 87
Skin disease 159, 160
Snake bite 20, 28, 37, 40, 46, 59, 64, 161, 162
Sneezing 81, 90, 130
Spermatorrhoea 35, 159
Sprain 124, 159
Stomatitis 48
Stye 95
Suicide 79, 83
Sunstroke 35
Swelling 41, 42, 44, 54
Swelling of gum 123
Swelling throat 39, 45

T

Tension 88
Tetanus 124
Tiger bite 4
Tingling 83
Tonic 4
Tonsils 87
Toothache 9, 13, 123, 158
Trembling 78
Tuberculosis 58, 140, 162
Tumour 5, 9, 124
Typhoid 40, 146

U

Ulcer 5, 160
Uric acid 126
Urinary problem 4
Urticariat 52

V

Vomiting 59, 79

W

Wasp sting 10
Weakness 19, 36

Whooping cough 50, 51, 90
Worm infection 20, 123
Wound 10, 52, 54, 56, 123, 140, 141, 146

Y

Yellow fever 145
Youthfulness 90

Z

Zoopharmacognosy 48